MW00843919

REEXAMINING THE QUANTUM–CLASSICAL RELATION

Beyond Reductionism and Pluralism

Classical mechanics and quantum mechanics are two of the most successful scientific theories ever discovered, and yet how they can describe the same world is far from clear: one theory is deterministic, the other indeterministic; one theory describes a world in which chaos is pervasive, the other a world in which chaos is absent. Focusing on the exciting field of "quantum chaos," this book reveals that there is a subtle and complex relation between classical and quantum mechanics. It challenges the received view that classical and quantum mechanics are incommensurable, and revives another largely forgotten tradition due to Niels Bohr and Paul Dirac. By artfully weaving together considerations from the history of science, philosophy of science, and contemporary physics, this book offers a new way of thinking about intertheory relations and scientific explanation. It will be of particular interest to historians and philosophers of science, philosophically inclined physicists, and interested non-specialists.

Alisa Bokulich received her Ph.D. in the History and Philosophy of Science from the University of Notre Dame in 2001. Her research focuses on the history and philosophy of physics, as well as on broader issues in the philosophy of science. In 2003 she was the recipient of a National Science Foundation Scholars Award, which supported much of her research for this book. She is currently a Professor in the Philosophy Department at Boston University and an active member of Boston University's Center for Philosophy and History of Science.

REEXAMINING THE QUANTUM–CLASSICAL RELATION

Beyond Reductionism and Pluralism

ALISA BOKULICH

Boston University

CAMBRIDGE UNIVERSITY PRESS
Cambridge, New York, Melbourne, Madrid, Cape Town, Singapore, São Paulo, Delhi

Cambridge University Press
The Edinburgh Building, Cambridge CB2 8RU, UK

Published in the United States of America by Cambridge University Press, New York

www.cambridge.org
Information on this title: www.cambridge.org/9780521857208

First published 2008

Printed in the United Kingdom at the University Press, Cambridge

A catalog record for this publication is available from the British Library

Library of Congress Cataloging in Publication Data
Bokulich, Alisa.
Reexamining the quantum-classical relation : beyond reductionism
and pluralism / Alisa Bokulich.
p. cm.
ISBN 978-0-521-85720-8
1. Quantum theory. 2. Physics–Philosophy. 3. Quantum theory–History.
4. Quantum theory–Philosophy. I. Title.
QC174.12.B65 2008
531–dc22
2008018902

ISBN 978-0-521-85720-8 hardback

For my mother
Elizabeth H. Payson
(1944–2006)
And for my teacher
James T. Cushing
(1937–2002)

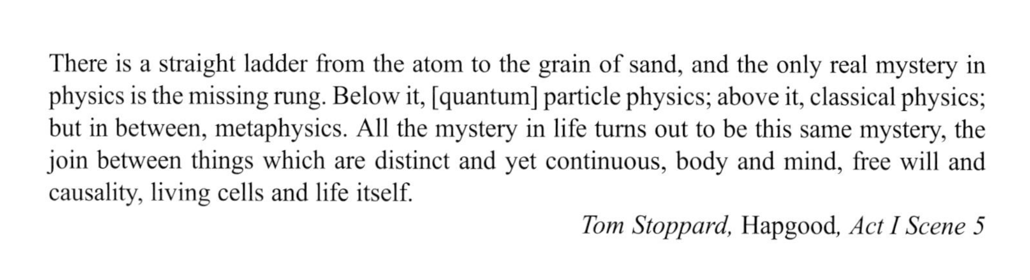

There is a straight ladder from the atom to the grain of sand, and the only real mystery in physics is the missing rung. Below it, [quantum] particle physics; above it, classical physics; but in between, metaphysics. All the mystery in life turns out to be this same mystery, the join between things which are distinct and yet continuous, body and mind, free will and causality, living cells and life itself.

Tom Stoppard, Hapgood, *Act I Scene 5*

Contents

Acknowledgements

In writing this book, I have benefited greatly from many physicists, historians, and philosophers generously sharing their time and expertise with me. I would especially like to thank Michael Berry and Bob Batterman for very helpful comments on Chapter 1, Mélanie Frappier and Michael Dickson for constructive comments on an earlier incarnation of Chapter 2, and Sam Schweber for insightful comments on Chapter 3. Chapter 4 first grew out of conversations with Peter Bokulich, and I am particularly grateful to Olivier Darrigol for enlightening discussions about Bohr's correspondence principle. Chapter 5 could never have been written without the invaluable conversations I had with Rick Heller and Dan Kleppner, and I am especially grateful to Steve Tomsovic and Rodolfo Jalabert for their careful reading of this chapter and detailed feedback. I would also like to thank Ernan McMullin and for Gordon Belot for their insightful comments on an earlier version of Chapter 6. Thanks are also owed to my colleague John Stachel, for reading various sections and for making BU a stimulating and congenial place to do history and philosophy of physics. Although I am indebted to these scholars for their feedback, they are in no way responsible for any remaining errors or idiosyncratic interpretations found here. Finally, I am very grateful to Jens Nöckel for generously providing me with the exquisite "bow-tie laser" image for the cover of this book; as Dirac might say, the science behind this image is as beautiful as the image itself.

The research for this book was made possible by a generous grant from the National Science Foundation, grant SES-0240328; I am extremely thankful for NSF's continued and invaluable support of interdisciplinary projects such as this. I would also like to thank Boston University's Humanities Foundation for financial support while part of this book was completed.

I am grateful to Michael Leach and Marina Werbeloff of Harvard University's Libraries for assistance with the Archive for the History of Quantum Physics, and for giving me permission to quote from it. I would also like to thank Sharon Schwerzel, head of the Paul A. M. Dirac Science Library at Florida State University, for providing me with copies of an unpublished lecture of Dirac's from the Paul A.M. Dirac Collection and for granting me permission to quote from it here. I am grateful to Frances Whistler of Boston University's Editorial Institute for her kind support in transcribing this lecture of Dirac's.

Many of the ideas presented here grew out of earlier journal articles. Some material from Chapters 2 and 3 comes with permission from "Open or closed? Dirac, Heisenberg, and the relation between classical and quantum mechanics," *Studies in History and Philosophy of Modern Physics* 35 (3): 377–96 (2004), copyright 2004 Elsevier Ltd. Sections 4.3 through 4.5 are based on my paper coauthored with Peter Bokulich: "Niels Bohr's generalization of classical mechanics," *Foundations of Physics* 35 (3): 347–71 (2005), copyright 2005 and provided with the kind permission of Springer Science and Business Media. The material connecting Heisenberg and Kuhn for Chapter 2 is based on "Heisenberg meets Kuhn: Closed theories and paradigms," *Philosophy of Science* 73: 90–107 (2006), copyright 2006 by the Philosophy of Science Association. Section 3.5 is from "Paul Dirac and the Einstein-Bohr debate," *Perspectives on Science* 16 (1): 103–14 (2008), copyright 2008 by the Massachusetts Institute of Technology. Finally, Section 5.5, Section 6.2, and Sections 6.4–6.6 draw on "Can classical structures explain quantum phenomena?" *British Journal for the Philosophy of Science*: 59 (2): 217–35 (2008).

I would also like to express my deep gratitude to Simon Capelin (the physical science publishing director at Cambridge University Press) for supporting this project from the very beginning, and to Graham Hart (the new philosophy and foundations of physics editor at Cambridge) for seeing this book through to completion. It has been a pleasure working with the entire CUP team. Thanks are also owed to my graduate students John Tietze, for help compiling the index, and Carolyn Suchy-Dicey, for help correcting the proofs.

Finally, on a personal note, I would like to thank my son, Julian, whose birth halfway through the writing of this book has provided me with many happy distractions. I am also grateful to my stepfather, Doug, for his support – especially in this last difficult year. My greatest debt, however, is owed to Peter Bokulich, who, by rarely agreeing with me, was a stimulating interlocutor for many of the ideas developed here. If it were not for his unwavering support and help in countless ways, this book would never have been completed. He has my eternal love and gratitude.

Introduction

There are many important questions that do not fall neatly into any one discipline; rather, their full investigation requires the integration of two or more distinct fields. This book is about just such a question – one that arises at the intersection of physics, philosophy, and history. The question can be simply stated as "What is the relation between classical and quantum mechanics?" The simplicity of the question, however, belies the complexity of the answer. Classical mechanics and quantum mechanics are two of the most successful scientific theories ever developed, and yet how these two very different theories can successfully describe one and the same world – the world we live in – is far from clear. One theory is deterministic, the other indeterministic; one theory describes a world in which chaotic behavior is pervasive, and the other a world in which it is almost entirely absent. Did quantum mechanics simply replace classical mechanics as the new universal theory? Do they each describe their own distinct domains of phenomena? Or is one theory really just a continuation of the other?

In the philosophy literature, this sort of issue is known as the problem of intertheoretic relations.[1] Currently, there are two accepted philosophical frameworks for thinking about intertheoric relations: the first is reductionism, and the second, pluralism. As we shall see, these labels each actually describe a family of related views. Reductionism is roughly the view that one theory can be derived from another, either by means of a logical deduction or the mathematical limit of some parameter. Theoretical pluralism, by contrast, takes each scientific theory to have its own distinct domain of laws, entities, and concepts, which cannot be reduced to those of any other theory. The central thesis of this book is that neither reductionism nor pluralism adequately describes the relation between quantum and classical mechanics.

[1] In the philosophical literature, as in the remainder of this book, the terms 'intertheoretic' and 'intertheory' are used interchangeably.

In searching for a new philosophical framework for thinking about intertheory relations, I turn to the history of science, and examine the philosophical views of three of the founders of quantum theory: Werner Heisenberg, Paul Dirac, and Niels Bohr. Perhaps surprisingly, all three of these figures accorded to classical mechanics a role of continued *theoretical* importance; none of them took classical mechanics to be a discarded theory, rendered useless for all but "engineering" purposes. Moreover, I shall argue that none of them took the relation between classical and quantum mechanics to be captured by the usual reductionist account in terms of the classical limit ($\hbar \rightarrow 0$). Despite these two important similarities, however, all three of them held a very different view of the quantum–classical relation. As we shall see, Heisenberg's account of classical and quantum mechanics as "closed theories" led him to adopt a version of theoretical pluralism; Dirac, by contrast, saw a deep analogy or structural continuity between these theories; and Bohr viewed quantum theory as a "rational generalization" of classical mechanics. I shall show that not only do these historical views have many interesting parallels with contemporary debates in the philosophy of science, but they can also suggest new ways in which our present-day debates might be moved forward.

The question of the relation between classical and quantum mechanics cannot be decided on purely historical or philosophical grounds, but also requires delving into contemporary research in physics. I shall focus on an area of scientific research known as semiclassical mechanics. Very roughly, semiclassical mechanics can be thought of as the study of "mesoscopic" systems that are in the overlap between the classically described macroworld and the quantum mechanically described microworld. As such, it is a field ideally suited for exploring questions about the relationship between classical and quantum mechanics. More specifically, I shall focus on a subfield known as quantum chaos. The name "quantum chaos" is something of a misnomer, since quantum systems cannot exhibit the sort of sensitive dependence on initial conditions characteristic of classically chaotic behavior.[2] Instead, the field of quantum chaos is concerned with the study of quantum systems whose classical counterparts are chaotic. As we shall see, these quantum-chaotic systems pose a number of unique challenges for an adequate characterization of the quantum–classical relation. At the heart of this book is a summary of four areas of research in semiclassical mechanics that involve quantum systems whose classical counterparts are chaotic. These are the semiclassical solution of the helium atom, diamagnetic Rydberg atoms, wavefunction scarring, and quantum dots. These case studies will function as the "data" against which the adequacy of the philosophical accounts of intertheory relations will be tested.

[2] For a review of the reasons why there cannot typically be chaotic behavior in quantum systems see Bokulich (2001, p. x).

I shall argue that there are three surprising lessons to draw from this examination of semiclassical research: First, there is a variety of *quantum* phenomena ranging from atomic physics to condensed-matter physics, for which semiclassical mechanics – not pure quantum mechanics – provides the appropriate theoretical framework for investigating, calculating, and *explaining* these phenomena. Second, these semiclassical methods and explanations involve a thorough hybridization of classical and quantum ideas. Far from being incommensurable theoretical concepts, they can be combined in both empirically adequate and conceptually fruitful ways. Finally, the classical structures (such as periodic orbits) appealed to in semiclassical mechanics are not simply useful calculational devices, but are actually manifesting themselves in surprising ways in quantum *experiments* (that is, in ways that are not simply the quantum behavior mimicking the classical behavior). This speaks to a much richer continuity of dynamical structure across classical and quantum mechanics than is usually recognized.

These features of semiclassical research pose two important challenges for contemporary philosophy of science. First, the semiclassical appeals to classical structures in explaining quantum phenomena do not fit easily with either of the current orthodox accounts of scientific explanation (that is, they are neither deductive–nomological nor causal explanations). I shall argue that a new philosophical account of scientific explanation is called for and outline what such an account might look like.

Second, this semiclassical research also poses a challenge to the adequacy of our current philosophical frameworks for thinking about intertheory relations. More specifically, it reveals that an adequate account of the relation between classical and quantum mechanics should not just be concerned with the narrow (though of course important) question of how to recover classical behavior from quantum mechanics, but rather should recognize the many structural correspondences and continuities between the two theories. Of the three historical views that I examine, I shall argue that Dirac's "structural continuity" view provides the most adequate foundation for a new philosophical account of intertheory relations – one that can incorporate these insights from semiclassical research. This new view, which I call interstructuralism, takes from theoretical pluralism the insight that predecessor theories such as classical mechanics are still playing an important theoretical role in scientific research; that is, quantum mechanics – without classical mechanics – gives us an incomplete picture of our world. From reductionism, however, it takes the lesson that we cannot rest content with the view that each of these theories describes its own distinct domain of phenomena. We stand to miss out on many important scientific discoveries and insights if we do not try to bring our various theoretical descriptions of the world closer together.

1

Intertheoretic relations: Are imperialism and isolationism our only options?

... that was to this, Hyperion to a satyr.

Shakespeare, Hamlet, *Act 1 Scene 2*

1.1 Introduction

The issue of intertheoretic relations is concerned with how our various theoretical descriptions of the world are supposed to fit together. As the physicist Sir Michael Berry describes it, "Our scientific understanding of the world is a patchwork of vast scope; it covers the intricate chemistry of life, the sociology of animal communities, the gigantic wheeling galaxies, and the dances of elusive elementary particles. But it is a patchwork nevertheless, and the different areas do not fit well together" (Berry 2001, p. 41). This uncomfortable patchwork exists even if we restrict our attention to within the field of physics alone. Physics itself consists of many subtheories, such as quantum field theory, quantum mechanics, condensed-matter theory, thermodynamics, classical mechanics, and the special and general theories of relativity – just to name a few. Each of these theories is taken to be an accurate description of some domain of phenomena, and insofar as they are supposed to be describing one and the same world, it is important to ask how these very different – and in many cases prima facie mutually inconsistent – theories are supposed to fit together.

Hitherto, the philosophical frameworks available for thinking about intertheory relations have been rather limited. Traditionally, discussions of intertheoretic relations have been framed in terms of reductionism.[1] The relationship between two scientific theories is taken to be either an accomplished, or an in-principle (though perhaps not in-practice) accomplishable, reduction of the higher-level (or predecessor)

[1] Throughout this book, the criticisms I raise against reductionism should be understood as being against *theory reductionism* or *explanatory reductionism*. At no point am I challenging ontological reductionism, or what philosophers sometimes call materialism. So, for example, I do not think that there are emergent properties that are not just the result of fundamental physical properties, their organization, and complex interactions.

theory to the lower-level (or successor) theory. On the reductionist picture, physics (or more precisely, high-energy physics) is the most fundamental and accurate description of the world, and the special sciences such as chemistry, biology, and psychology are merely incomplete shadows of – or approximations to – this more fundamental level of description. So on this view, for example, as biological theories become better and better, they should become more and more indistinguishable from theories in physics. Reductionism has furthermore been used as a justification for why more research funds should be invested in high-energy physics than in other areas of science, much to the vexation of scientists in other fields.[2]

In recent years, however, reductionism has increasingly fallen into disrepute. Reductionists are now likened to imperialists, who aim to illegitimately extend the power and dominion of a particular scientific theory by direct territorial acquisition. In place of reductionism, the new orthodoxy has become theoretical pluralism. Pluralists, by contrast, allot to each scientific theory its own circumscribed domain, not to be infringed upon by any of its neighbors. On this view, each of the special sciences has its own entities, laws, and descriptions, none of which are any less fundamental than the entities, laws and descriptions of physics. The theoretical pluralist Jerry Fodor, for example, titles his more recent (1997) defense of anti-reductionism "Special sciences: Still autonomous after all these years."[3]

In trying to avoid the reductionist's imperialism, however, the pluralists have adopted a position that is dangerously close to isolationism. In their renunciation of reductionism, theoretical pluralists have also renounced the important benefits that come from building strong and intimate ties with neighboring theories. Indeed it is arguably in trying to build bridges between these various different scientific domains that some of the most exciting new developments and discoveries are made.

In the following two sections I provide a brief overview of the various forms of reductionism and theoretical pluralism.[4] This will provide a useful framework for locating the alternative approaches to intertheoretic relations that I introduce in Chapters 2, 3, and 4. In Sections 1.4 and 1.5, I turn more specifically to the case of classical and quantum mechanics, and examine in some detail the most widely

[2] For example, Steven Weinberg in his chapter "Two Cheers for Reductionism" writes, "The reason we give the impression that we think that elementary particle physics is more fundamental than other branches of physics is because it is. I do not know how to defend the amounts being spent on particle physics without being frank about this" (Weinberg 1992, p. 55). In this book, Weinberg also briefly discusses how his views on reductionism have been challenged by the evolutionary biologist Ernst Mayr and the condensed-matter physicist Philip Anderson.

[3] While this paper is directed more specifically at Jaegwon Kim's reductionist arguments in psychology, it is a continuation of Fodor's (1974) arguments for the irreducibility and autonomy of the special sciences.

[4] Regrettably the brevity of this overview means I will be unable to do full justice to the subtleties of the various views summarized. As some compensation, I have tried to provide ample references for the interested reader to learn about these views in more depth.

received account of the relation between these theories, which is a form of reductionism. This will require a foray into contemporary research in physics that seeks to explain how quantum theory can recover (or reduce to) the everyday classical world we observe around us. The challenges raised there against the most widely received form of reductionism will be the first step in the argument, taken up again in Chapters 5, 6, and 7, that neither reductionism nor pluralism gives an adequate account of the relationship between classical and quantum mechanics.

1.2 Traditional accounts of reductionism

Part of the difficulty in deciding whether or not one theory is reducible to another is that there is no univocal understanding of what reductionism requires.[5] Reductionism can, for example, be construed as a thesis about ontologies, laws, theories, methodologies, or even linguistic expressions. Nor are these various construals of reductionism mutually exclusive. Furthermore, reductionism can be understood either as a synchronic relation – that is, a relation between two concurrent theories that belong to two different levels of description – or a diachronic relation describing the relation between a historical predecessor theory and its successor. This distinction between synchronic and diachronic reduction often becomes blurred in cases like classical and quantum mechanics, where the historical predecessor is a higher-level "macrotheory" and its successor is a "microtheory."

Reductionism is best thought of not as a single approach, but rather as a framework, or family of approaches. Different approaches to reductionism can be distinguished by the different ways in which the reductive relation is characterized. The three most prominent approaches to reductionism (which shall be briefly discussed in turn) are, first, Nagelian reductionism as logical deduction; second, Kemeny–Oppenheim reduction as an eliminative re-systematization; and finally, what is often called the "physicist's reductionism" as the asymptotic limit of some parameter.

The best-known formulation of reductionism in philosophy is that of Ernest Nagel. According to Nagel, "a reduction is effected when the experimental laws of the secondary science … are shown to be the logical consequences of the theoretical assumptions (inclusive of the coordinating definitions) of the primary science" (Nagel [1961] 1979, p. 352). Here, the secondary science is the predecessor theory (e.g., thermodynamics) and the primary science is the successor theory (e.g., statistical mechanics). Since reduction on this model is a logical derivation, there can be no term in the predecessor theory that is not in the successor theory. To overcome this difficulty, Nagel introduces the condition of connectability, which

[5] For a taxonomy and discussion of various forms of reductionism see, for example, Sarkar (1992).

requires bridge principles or laws that would connect different terms in the two theories.

Critics quickly pointed out a number of theoretical problems with Nagel's model of reduction. For example, the required bridge laws can rarely be found, and the reduced theory requires assumptions that, in light of the new theory, are strictly speaking false.[6] More generally the conception of a scientific theory as a system of laws, on which Nagel's account of reductionism rests, has also been called into question. For example, it is not clear that theories in the special sciences, such as biology, have laws at all.[7] Most damaging, however, is the fact that no actual cases of intertheory relations have been able to fit Nagel's model.[8]

The second traditional philosophical approach to reductionism is due to John Kemeny and Paul Oppenheim. Partly in response to difficulties that they see in Nagel's account, they offer an alternative model of reductionism, which emphasizes simplicity and conceptual economy. According to Kemeny and Oppenheim, a scientific theory is nothing but a systematization of all of our observations to date. They write, "In place of an infinite set of observation statements we are given a reasonably simple theory. Such a theory has the same explanatory ability as the long (or infinite) list of statements, but no one will deny that it is vastly simpler and hence preferable to such a list. Only thus do we see the need for introducing theoretical terms" (Kemeny and Oppenheim 1956, p. 12). It is on this remarkably impoverished account of scientific theories that Kemeny and Oppenheim base their notion of reduction: A predecessor theory, T_2, is reduced to a successor theory, T_1, if all observational data that T_2 can explain is also explainable by T_1. To this they add the further condition that T_1 must be at least as well systematized as T_2, where "systematized" means "simpler" unless the loss of simplicity is counterbalanced by an increase in the strength or scope of the successor theory.

An important consequence of Kemeny and Oppenheim's account of reduction is that the predecessor theory, which is now nothing more than a superfluous systematization of observation statements, can simply be eliminated. Theory reduction, on this model more closely resembles theory replacement. Kemeny and Oppenheim note that while Nagel attempts to establish a direct relation between the two theories, "[our] connection is indirect … [O]f course, each set of theoretical terms must be connected to observational terms, and hence to each other, but this connection is normally much weaker than a full translation" (Kemeny and Oppenheim 1956,

[6] For an overview of some of these early objections to Nagel's model, as well as a defense of a neo-Nagelian model of reduction that attempts to overcome these objections, see Schaffner (1967).

[7] See, for example, Kitcher (1984).

[8] The literature on this topic is vast. For a sampling of challenges to Nagel's reduction in the context of specific theory pairs, see Fodor (1974) for psychology to neuroscience, Kitcher (1984) for classical Mendelian genetics to molecular biology, Scerri (1994) for chemistry to quantum mechanics, and Sklar (1999) for thermodynamics to statistical mechanics.

p. 16). If one abandons the idea that there is a fixed set of observation statements that is preserved in the move from one theory to another, as many post-positivist philosophers of science do, then even this indirect relation between the predecessor and successor theory is lost.

In addition to the deductive and eliminative models of theory reduction, there is a third general approach to intertheoretic reduction that is typically found in the physical sciences. Thomas Nickles labels this "reduction$_2$," which he notes is "best described by 'inverting' the usual concept of reduction, so that successors are said to reduce to their predecessors (not vice versa) under limiting operations" (Nickles 1973, p. 181). As Nickles notes, what amounts to a limiting operation can vary widely. Typically a parameter (or combination of parameters) is allowed to go to some limit. For example, it is commonly said that the special theory of relativity reduces$_2$ to Newtonian dynamics in the limit of small velocities (that is $v^2/c^2 \rightarrow 0$). In the context of classical and quantum mechanics, this third approach to reductionism is typically referred to as the "classical limit." Since this approach is by far the most widely received account of the relation between classical and quantum mechanics, it will be examined in some detail in Section 1.4. Before turning to physicists' characterizations of the classical limit, however, we shall briefly examine the chief rival to these traditional accounts of reductionism, namely, theoretical pluralism.

1.3 Theoretical pluralism

Like reductionism, theoretical pluralism is best thought of as not a single view, but rather a family of approaches. Theoretical pluralism, which is sometimes referred to as "scientific pluralism," has also been defended under the rubric of the "disunity of science." As a negative thesis, pluralism denies that the world is such that it can be explained by a single unified set of fundamental principles or laws.[9] The plurality of scientific theories describing different domains is neither a temporary feature of science, nor a permanent feature that is merely a consequence of our epistemic limitations as human knowers. Rather, there is some sense in which nature itself demands this plurality of descriptions; hence, any approach that seeks to reduce or eliminate this plurality is methodologically misguided and will result in a misrepresentation of nature.

Three different subspecies of theoretical pluralism can be distinguished: type-I theoretical pluralism, which defends the necessity of multiple

[9] Some versions of pluralism simply remain agnostic about this question, seeing this agnosticism as sufficient to undermine the normative claim that scientists *ought* to be seeking a unified account, and that the acceptability or maturity of a theory is, in large part, to be measured by the extent to which it achieves that unity. See, for example, Kellert, Longino, and Waters (2006).

scientific theories or models in describing *different* domains of phenomena; type-II, which defends a plurality of theories or models in describing the *same* (single) domain of phenomena; or type-III theoretical pluralism, which argues for both. Type-I theoretical pluralism is perhaps the most common, and is typically invoked to protect the autonomy of the special sciences. Challenges facing this version of theoretical pluralism include the problem of consistency that arises at the interface, or borderlands, between our various scientific theories. For the second and third types of pluralism, this problem of inconsistency is even more troubling, in so far as it is no longer confined to these borderlands, but rather pervades these theories. Even more troubling, however, is that these latter two types of theoretical pluralism threaten to lead to a relativism of "anything goes." Hence, those who want to embrace pluralism without relativism are faced with the challenge of articulating a set of constraints on which models and theories are to be counted among the scientifically legitimate ones for that domain of phenomena (Kellert *et al.* 2006, p. xiii).

Historically, the thesis of incommensurability played a central role in the move away from reductionism and towards theoretical pluralism. In responding to purported cases of reduction, such as the reduction of special relativity to Newtonian dynamics, both Thomas Kuhn ([1962] 1996) and Paul Feyerabend (1962) argued that, although the predecessor and successor theories may use the same terms, such as "mass" and "space" – the meanings of these terms have fundamentally changed. This incommensurability of terms blocks any attempt to connect the two theories via bridge principles; hence, any claim to have reduced one theory to another is unfounded.

The thesis of incommensurability led Feyerabend to endorse a version of theoretical pluralism which he defines as "the simultaneous use of mutually inconsistent theories" to describe a single domain of phenomena (Feyerabend [1965] 1983, p. 149). On this view, scientists ought to invent and develop in detail as many alternatives to the currently accepted theory as possible. Hence, on the taxonomy we introduced above, Feyerabend would best be characterized as a type-II pluralist.

According to the early Feyerabend at least, theoretical pluralism is defended on the methodological grounds that it provides a more rigorous testing of our theories than simply comparing them with "the facts."[10] That is, in so far as these rival alternative theories will be incommensurable with the currently accepted theory, and incommensurable with one another, they will provide a more rigorous testing environment. On this point, however, the views of Feyerabend and Kuhn diverge: Kuhn does not advocate theoretical pluralism as a central methodology of science – except

[10] In so far as Feyerabend's theoretical pluralism is defended on methodological – rather than metaphysical – grounds, it is compatible with the view that there is in principle one correct unified description of the world.

perhaps in periods of so-called crisis. Instead, Kuhn's view remains by and large closer to the eliminative model of reductionism as straightforward theory replacement, or what I have elsewhere referred to as a "serial theoretical monogamy."[11]

One of the clearest contemporary articulations of theoretical pluralism is found in the work of Nancy Cartwright (1995; 1999). In her book *The Dappled World* she writes, "[T]he theory is successful in its domain … Theories are successful where they are successful, and that is that. If we insist on turning this into a metaphysical doctrine, I suppose it will look like metaphysical pluralism" (Cartwright 1999, p. 31). Cartwright defends this pluralism on metaphysical, rather than simply epistemological, grounds. More specifically, she defines her metaphysical version of this thesis as the claim that "nature is governed in different domains by different systems of laws not necessarily related to each other in any systematic or uniform way" (Cartwright 1999, p. 31). This view undermines the reductionist program at its very foundation by denying that there is any theory or set of laws that is universally valid. Thus Cartwright's view seems best described as a version of type-I theoretical pluralism, in so far as she seems to suggest that once a domain of phenomena is delimited, then there is one correct theoretical description of that domain. In other words, her challenge is specifically to the *universality* of physical theories or models, not to their correctness within some circumscribed domain.

Others, such as John Dupré (1995; 1996a; 1996b), have defended a similar thesis under the rubric of "the disunity of science." Dupré, like Cartwright, seeks to defend pluralism on metaphysical grounds: "[T]he picture of science as radically fractured and disunified has a role for metaphysics, and moreover that … set of metaphysical views is entirely plausible" (Dupré 1996b, p. 101). As a philosopher of biology, Dupré's argument for pluralism rests not on an analysis of laws, but rather on a rejection of the following two assumptions that he sees at the foundation of reductionism: first the assumption that there is a unique taxonomy of kinds, and, second, the assumption of causal completeness.[12] According to Dupré, the problem is not that the taxonomies that science provides fail to capture real kinds out there in the world; he thinks they do. Rather, the problem is that these taxonomies are not unique. Dupré argues that one and the same entity can legitimately belong to several different natural kinds, a view he describes as "promiscuous realism."

[11] This phrase was introduced in Bokulich (2006, p. 105). Unlike Kemmeny and Oppenheim's eliminativism, however, Kuhn denies that the observations remain unchanged in the move from one paradigm to another (Kuhn [1962] 1996, pp. 134–5). Although the incommensurability thesis did not lead Kuhn to embrace theoretical pluralism, I shall argue in some detail in the next chapter, that an early formulation of the incommensurability thesis did play a central role in leading the physicist Werner Heisenberg to embrace a version of type-I theoretical pluralism in his account of the relation between classical and quantum mechanics.

[12] Dupré defines causal completeness as "the assumption that for every event there is a complete causal story to account for its occurrence" (Dupré 1996a, p. 99). Regrettably, I will be unable to take the time to discuss this argument here.

He sees this pluralism as undermining (Nagelian) reductionism insofar as "the individuals that would have to be assumed for the derivation of macrotheory cannot be identified with those that are the subjects of descriptive accounts at the next-lower level" (Dupré 1996a, p. 116). For example, in his defense of the claim that ecology cannot be reduced to physiology, he introduces the case of the lynx and the hare, and notes that "the ideal hare that the physiologist might construct . . . is just not the same as the ideal hare that is hunted by the ecologist's ideal lynx" (Dupré 1996a, p. 118).[13] It is this inability to identify theoretical terms in different sciences that secures the autonomy of ecology from any imperialist ambitions of physiology.

It seems that Dupré's defense of the disunity of science is best characterized as a version of type-III theoretical pluralism. On the one hand, Dupré is concerned with securing the autonomy of the special sciences by arguing for their irreducibly distinct domains of application, and hence endorses a type-I pluralism. On the other hand, he also wants to challenge the view that within a single domain – or even for a single entity – there is one theory that best describes that domain or entity. Hence, he endorses a type-II pluralism as well.

In addition to these defenses of theoretical pluralism in the contexts of physics and biology, pluralism is also frequently defended in the context of psychology. For example, in his seminal paper "Special sciences (or: The disunity of science as a working hypothesis)," Jerry Fodor has argued that psychology cannot be reduced to the neurosciences.[14] Like the other theoretical pluralists we have discussed so far, he attributes this irreducibility to the way the world is, rather than to our epistemic limitations:

> The existence of psychology depends not on the fact that neurons are so sadly small, but rather on the fact that neurology does not posit the natural kinds that psychology requires. I am suggesting, roughly, that there are special sciences not because of the nature of our epistemic relation to the world, but because of the way the world is put together.
>
> *(Fodor 1974, p. 113)*

Hence psychology is – and will always remain – an autonomous science. Unlike Dupré's, however, Fodor's pluralism of natural kinds seems to be confined to distinct scientific theories, and hence is best described as a version of type-I theoretical pluralism.[15]

What begins to emerge from these defenses of theoretical pluralism is a view of science as "radically fractured," with distinct scientific communities whose respective members have difficulty communicating with each other. The image that

[13] The rejection of causal completeness part of the lynx and hare story is described in Dupré (1996a, p. 117).

[14] Fodor's title is, of course, a play on Oppenheim and Putnam's classic 1958 defense of reductionism "Unity of Science as a Working Hypothesis."

[15] Type-II pluralism by contrast would advocate the development of multiple rival *psychological* theories to describe the same psychological phenomena.

Cartwright uses to illustrate her pluralism is one in which scientific theories are likened to isolated balloons – tethered to the empirical world – but not intimately interconnected with each other.[16] If each theory has its own proper domain, then it would seem that the search for intertheoretic relations and interdependencies is fundamentally misguided.

A small step towards ameliorating this isolationist tendency of theoretical pluralism is taken by Peter Galison (1997). Like Cartwright, Dupré, and Fodor, Galison is a firm defender of the disunity of science. Rather than focusing on theoretical disunity, however, Galison is primarily concerned with the disunity between what he calls the theoretical, experimental, and engineering subcultures of physics. Regardless of whether one is concerned with theoretical disunity or a disunity of subcultures, as Galison notes, the isolationist picture is simply not empirically adequate. Scientists from different disciplines do communicate, interact, and collaborate with one another. As a way of trying to reconcile this observation with pluralism Galison introduces his notion of border trading zones. He explains,

> Two groups can agree on rules of exchange even if they ascribe utterly different significance to the objects being exchanged; they may even disagree on the meaning of the exchange process itself. Nonetheless, the trading partners can hammer out a *local* coordination despite vast *global* differences … [T]he site at which the local coordination between beliefs and actions takes place … is a domain I call the trading zone.
>
> *(Galison 1997, pp. 783–4)*

According to Galison, problems of linguistic incommensurability can be overcome through the development of pidgins – partial contact languages constructed from elements of two full languages. While Galison's view is a step in the right direction, the interactions and connections it allows for are still, by and large, confined to the border regions. When it comes to the heartlands of these theories, isolationism still prevails.

Returning to the case of classical and quantum mechanics, both Cartwright (1999) and Gordon Belot (2000) have argued that the relation between these theories is best characterized in terms of theoretical pluralism. Cartwright, for example, writes,

> [I]t is generally assumed that we have discovered that quantum mechanics is true … and hence classical mechanics is false … All evidence points to the conclusion that … Nature is not reductive and single minded … and is happily running both classical and quantum mechanics side-by-side.
>
> *(Cartwright 1995, p. 361)*

[16] For the balloon metaphor see Cartwright (1995; 1999).

Similarly, Belot provocatively asks,

> What grounds do you have for believing that quantum mechanics provides, in principle if not in practice, a *more accurate* view of the world than does classical mechanics? Granting that the two theories diverge in their empirical predictions, it can only be that quantum mechanics has proved to be a far better way of approaching the microworld. But what grounds can we have for confidence that this superiority extends to the macroworld?
>
> *(Belot 2000, p. S464)*

The divergence of predictions that Belot refers to here arises in the context of the quantum description of systems that are classically chaotic, and will be described in more detail in the next section. While Belot does not defend a strong metaphysical thesis of theoretical pluralism the way Cartwright does, the implication of the above passage is that quantum mechanics may not be a universal theory; rather, it may be that both classical and quantum mechanics are required for a correct description of the world, each having its own limited and irreducible domain of applicability.[17]

The isolationist tendencies of theoretical pluralism are fueled when one notes that classical and quantum mechanics are typically formulated in radically different mathematical frameworks. To follow the incommensurability metaphor, one might say that the quantum and classical theorists, though they both use the words "state" and "momentum," are really speaking different languages. Whereas classically, observables are represented as functions on phase space and the state of the system can be represented by a point, in quantum mechanics observables are represented by operators on Hilbert space and the (pure) state of the system is represented by a vector in that complex vector space. These striking differences can be somewhat ameliorated by noting that there are Hilbert space formulations of classical mechanics (such as the one introduced by Bernard Koopman in 1931) and phase space formulations of quantum mechanics (such as the one introduced by Eugene Wigner in 1932). Despite these translations, however, there will remain some irreducible differences, such as the fact that the algebra of quantum mechanics is fundamentally noncommutative, while that of classical mechanics is not.[18]

As we noted at the outset, the debate between reductionism and pluralism must be well-grounded in the scientific details of the theories in question. Hence, in order to adequately assess which account of intertheory relations best describes the relation between classical and quantum mechanics, we must first take a closer look at the state of current research in physics to see whether physicists have fared any better in effecting a reduction between these theories.

[17] Belot hedges his bets by noting that it is possible that there may be a "fundamentalist" (i.e., quantum-mechanical) solution to the quantum chaos crisis (Belot 2000, p. S462).

[18] Heisenberg ([1925] 1967) and Dirac (1925).

1.4 The limits of the classical limit

It is typically argued that Nickles' reduction$_2$ model best describes the relationship between classical and quantum mechanics. On this view, recall, the reduction relation is characterized in terms of the mathematical limit of a parameter. The most common formulation of this limit is in terms of Planck's constant going to zero, that is, $\hbar \to 0$. This statement is somewhat misleading, however, since Planck's constant is, of course, a *constant*, and cannot take on any value other than the one it has (1.05457×10^{-34} Joule-seconds).[19] What is actually meant by $\hbar \to 0$ is that one considers the limit of a dimensionless quantity formed by the ratio of Planck's constant to some other quantity with the same dimensions. What this quantity will be depends on the particular system of interest.

The historian of science Max Jammer identifies Max Planck's 1906 book *The Theory of Heat Radiation* as the first place where this limit was formulated (Jammer 1966, p. 109). The example of how Planck's (quantum) radiation formula reduces to the (classical) Rayleigh–Jeans formula provides a nice illustration of how the $\hbar \to 0$ characterization of the classical limit is supposed to work. Planck's formula for the energy of blackbody radiation at frequency ν is

$$U(\nu) = \frac{8\pi V}{c^3} h\nu^3 \frac{1}{e^{h\nu/kT} - 1}, \tag{1.1}$$

where V is the volume of the cavity, c is the speed of light, k is Boltzmann's constant, and T is the temperature. One cannot, of course, simply set Planck's constant, h, to zero in this expression and expect to recover the classical Rayleigh–Jeans formula for the energy density, since doing so would make the entire expression disappear. Instead one needs to find the appropriate dimensionless quantity with Planck's constant in the numerator. In the case of this particular physical system, the quantity we are interested in taking the limit of is the dimensionless $h\nu/kT$. In the limit where $h\nu/kT$ is very small, the exponential can be expanded and only the first two terms, $1 + h\nu/kT$, kept. Inserting this back into Planck's formula, we get

$$U(\nu) = \frac{8\pi V}{c^3} kT\nu^2, \tag{1.2}$$

which is the classical Rayleigh-Jeans formula, no longer containing Planck's constant. With regard to this example, Jammer quotes Planck as saying, "The classical theory can simply be characterized by the fact that the quantum of action becomes

[19] As the bar indicates, this quantity is strictly speaking Planck's constant h divided by 2π.

infinitesimally small" (Jammer 1966, p. 109 citing Planck 1906, p. 143).[20] This quotation is potentially misleading, however, in that it suggests this limit recovers the classical regime for *all* physical systems. One must be careful in concluding from specific concrete examples such as this, that therefore classical mechanics – in its entirety – is "simply" recovered from quantum mechanics in this limit.

As we have already noted, the nature of the $\hbar \to 0$ limit depends on the particular system of interest, and for the vast majority of physical systems, this limit is singular. A singular limit is one in which the nature of the solution *near* the limit is fundamentally different from the nature of the solution *at* the limit. Berry graphically illustrates the character and implications of singular limits as follows:

> Biting into an apple and finding a maggot is unpleasant enough, but finding half a maggot is worse. Discovering one-third of a maggot would be more distressing still: The less you find, the more you might have eaten. Extrapolating to the limit, an encounter with no maggot at all should be the ultimate bad-apple experience.
>
> *(Berry 2002, p. 10)*

As Berry explains, a very small maggot fraction is qualitatively quite different from no maggot at all, hence the character of the solution near the limit is not always a reliable guide to the character of the solution at the limit.

The challenge posed by the typically singular nature of the $\hbar \to 0$ limit can be seen in the following well-known example of a particle incident on a potential barrier.[21] Consider a particle of energy E and a potential step barrier of height $V(x)$, where $E > V(x)$. Classically, the particle should sail through the barrier; there will be no reflection. Quantum mechanically, however, there is a small probability that the particle will be reflected by the barrier (a so-called reflection *above* a barrier). So in addition to the incident wave, there will also be a reflected wave and a transmitted wave, and the wavefunction will take the following form:

$$\begin{aligned} \psi(x) &= \exp\left(\frac{ip_1 x}{\hbar}\right) + R\exp\left(\frac{-ip_1 x}{\hbar}\right) \quad (x \ll 0) \\ &= T\exp\left(\frac{ip_2 x}{\hbar}\right) \quad (x \gg 0) \end{aligned} \tag{1.3}$$

[20] I must admit that I am unable to find this quotation either in the original German first edition or in Dover's 1959 English translation of the second edition of Planck's *Theory of Heat Radiation*. Following Jammer, this quotation has now become widely repeated (appearing, for example, in Liboff's (1984) *Physics Today* article) and is thought to be a generic description of the quantum–classical relation, not just specific to this context. What one does readily find in Planck's book is the much more modest claim that "for a vanishingly small value of the quantity element of action, h, the general formula [Planck's formula] degenerates into *Rayleigh's* formula" (Planck 1959, p. 170).

[21] This example follows Berry and Mount (1972, p. 318), where further details can be found.

where R and T are the reflection and transmission amplitude coefficients, and p_1 and p_2 are the momenta in the two regions of constant potential. Consider the reflection coefficient $|R^2|$: Classically, this reflection coefficient should be exactly zero, corresponding to the fact that in a classical world there is no reflection of the particle by the barrier. The dimensionless quantity we are interested in taking the limit of for this particular physical system is $\hbar/p_2L$, where L is a measure of the smoothness of the step. If the step is smooth, then, in the "semiclassical" regime where $(\hbar/p_2L) \rightarrow 0$, the reflection coefficient takes the form

$$\left|R^2\right| = \exp(-4\pi p_2 L/\hbar). \tag{1.4}$$

The difficulty, as Berry and Mount explain, is that "considered as a function of $\hbar$, the semiclassical reflection coefficient does not vanish as some simple power in the classical limit [at $\hbar = 0$], but has an essential singularity" (Berry and Mount 1972, pp. 318–19).[22] In other words, physically it is very important *how* the reflection coefficient vanishes as $\hbar \rightarrow 0$, not just that it vanishes for $\hbar = 0$. If the limit had not been singular, then it would have been possible to express the reflection coefficient as a power series in $\hbar$, with the first term proportional to $\hbar$ rather than exponentially small.[23]

More generally, if the classical limit were captured by $\hbar \rightarrow 0$, then one would expect to be able to write down quantum quantities as a convergent series in Planck's constant, with the first term being the classical quantity. Schematically, if G is the more general theory (e.g., quantum mechanics), S the more special one (e.g., classical mechanics), and δ the dimensionless parameter, then if $G \rightarrow S$ as $\delta \rightarrow 0$, one should be able to write G as a Taylor expansion in δ about S as $G = S + \delta\, S_1 + \delta^2\, S_2 + \cdots$ (Berry 1991, p. 256). The fact that this limit is typically singular, however, means that this is not possible. This has led both Berry (1991; 1994) and Robert Batterman (1995; 2002) to conclude that the singular nature of the $\hbar \rightarrow 0$ limit prohibits a straightforward reduction of quantum mechanics to classical mechanics.

Further problems arise for the $\hbar \rightarrow 0$ characterization of the classical limit when one considers systems that can exhibit classical chaos. Since the structures of classical chaos emerge in the long-time limit ($t \rightarrow \infty$), finding the classical limit of a chaotic system, requires taking both the $t \rightarrow \infty$ and $\hbar \rightarrow 0$ limits. The difficulty,

[22] An "essential singularity," recall, is defined as one near which the function exhibits extreme behavior. On a terminological note, Berry refers to $\hbar \rightarrow 0$ as the "semiclassical limit", reserving the name "classical limit" for the actual limiting value (Berry and Mount 1972, p. 319).

[23] The situation is even more troubling for a discontinuous step, where the reflection coefficient is independent of $\hbar$, and hence does not go to zero as $\hbar \rightarrow 0$ (Berry personal communication). For all smooth potentials, by contrast, the reflection coefficient does disappear for $\hbar = 0$, but the manner in which it disappears depends in an interesting way on the degree of smoothness of the potential. See Berry (1982) for further details.

however, is that these two limits do not commute, which means that taking the classical limit $\hbar \rightarrow 0$ first and then the long-time limit $t \rightarrow \infty$ does not yield the same result as taking the long-time limit first and the classical limit second. This means not only that it is nontrivial to obtain the correct classical limit of a quantum system whose classical counterpart is chaotic, but also that the physical behavior described in this combined limit can be surprisingly complex (Berry 1988). The challenges – and opportunities – that chaos poses for understanding the quantum–classical relation will be a recurring theme of this book.

In response to these difficulties, one might be tempted to conclude that one has simply picked the wrong parameter for characterizing the classical limit. An alternative candidate, that has been around almost as long as Planck's limit, is the limit of large quantum numbers, expressed as $n \rightarrow \infty$. Although not identical to it, this $n \rightarrow \infty$ limit is typically associated with Niels Bohr's correspondence principle, which shall be discussed in detail in Chapter 4. Quantum numbers describe quantities, such as energy and angular momentum, that are conserved by the quantum dynamics. One of the characteristic features of quantum systems is that certain observables, such as energy, have a discrete spectrum of eigenvalues. The quantum number n is typically referred to as the principal quantum number, and denotes the energy eigenvalues. In a hydrogen atom, for example, this number depends on the distance of the electron from the nucleus. As n becomes large, the spacings between adjacent energy eigenvalues become closer and closer together. Intuitively, the classically expected continuum of possible energy levels is reached when the spacing between energy eigenvalues becomes negligible, that is, in the limit of large n.

As several authors have pointed out, however, the limit of large quantum numbers is not sufficient for characterizing the classical limit. Albert Messiah, for example, in his classic textbook on quantum mechanics argues that while the negligible spacing between eigenvalues characterized by $n \rightarrow \infty$ is a *necessary* condition for the recovery of classical behavior, it is not a *sufficient* condition:

> In order that the approximation [of classical mechanics] be justified, it is necessary that this spacing could be considered negligible; that is the case if *large quantum numbers* are involved ... The condition is certainly not sufficient; thus, some purely quantum-mechanical effects such as the uncertainty reltions are not related to the discreteness of certain spectra.
>
> *(Messiah 1965, p. 214)*

His concern is that there are quantum phenomena that do not depend on n, and hence will not disappear in this limit.

More recently Richard Liboff (1984) has shown that even for quantum phenomena that *are* related to the discreteness of certain spectra the limit $n \rightarrow \infty$ does not always capture the classical limit. More specifically, he provides two simple counterexamples to the expectation that the quantum frequency spectrum of a periodic system approaches the classical spectrum in the limit of large n. The two systems he considers are a particle in a cubical box and a rigid rotator. In both cases it turns out that even in the $n \rightarrow \infty$ limit adjacent frequencies, $\nu_Q^{(n)}$ and $\nu_Q^{(n+1)}$, remain separated as $\nu_Q^{(n+1)} = \nu_Q^{(n)} + h/I$, where h is Planck's constant and I is a constant with dimensions of moment of inertia. In other words, the quantum frequency spectra remain discrete, rather than approximating the classical continuum.

In response to these difficulties in capturing the classical limit with the $\hbar \rightarrow 0$ and $n \rightarrow \infty$ limits individually, Hassoun and Kobe (1989) have proposed that an adequate characterization requires both limits to be taken simultaneously. They consider three well-known quantum-mechanical systems: the harmonic oscillator, the particle in a box, and the hydrogen atom. For these three systems they conclude:

> It is necessary to synthesize the Planck and Bohr formulations of the correspondence principle. Both formulations are used concurrently in the sense that the Planck constant goes to zero and the appropriate quantum number goes to infinity, subject to a constraint that their product be held fixed at the appropriate classical action. Meaningless results are obtained for the classical limit of quantum mechanical eigenvalues if one limit is taken without the other.
>
> *(Hassoun and Kobe 1989, p. 658)*[24]

For example, in the case of a harmonic oscillator, the energy eigenvalues obtained by solving the Schrödinger equation are

$$E_n = (n + \frac{1}{2})\hbar\omega \qquad n = 0, 1, 2, 3, ..., \tag{1.5}$$

where ω is the frequency. Classically the energy should be able to take on any arbitrary value. Hassoun and Kobe note that applying $\hbar \rightarrow 0$ alone leads to the energy going to zero for all n; alternatively, applying the $n \rightarrow \infty$ alone results in the energy going to infinity. If, however, both limits are taken such that the product is fixed at the classical action J, according to $nh = J$, then one obtains

$$E_n \rightarrow E_{\text{classical}} = \left(m\omega^2 A^2\right)/2, \tag{1.6}$$

the correct expression for the energy of a classical oscillator. Hence for these sorts of physical systems at least, while neither limit alone is adequate for characterizing the classical limit, both limits when taken in conjunction are adequate.

Another way in which physicists have sought to characterize the relation between classical and quantum mechanics is in terms of Ehrenfest's theorem. Intuitively,

[24] As I shall argue in Section 4.2, despite the ubiquity of this sort of interpretation of the correspondence principle in the physics literature, this is not in fact what Bohr meant by the correspondence principle.

Ehrenfest's theorem says that – under certain conditions – the mean values of the coordinates and momenta of a quantum system follow a classical trajectory. For example, consider a one-dimensional particle moving in a scalar potential $V(\mathrm{x})$ which generates a force $F(x) = -\nabla\ V(x)$. In the Heisenberg picture of quantum mechanics, the equations of motion for the position and momentum operators are

$$\frac{\mathrm{d}\hat{q}}{\mathrm{d}t} = \frac{\hat{p}}{m} \text{ and } F(\hat{q}) = \frac{\mathrm{d}\hat{p}}{\mathrm{d}t}. \tag{1.7}$$

Taking the average of this in some state yields Ehrenfest's theorem

$$\frac{\mathrm{d}\langle\hat{q}\rangle}{\mathrm{d}t} = \frac{\langle\hat{p}\rangle}{m} \text{ and } \langle F(\hat{q})\rangle = \frac{\mathrm{d}\langle\hat{p}\rangle}{\mathrm{d}t}. \tag{1.8}$$

If one can approximate the average of the function of position with the function of the average position – what we might call the "Ehrenfest substitution" – then one arrives at

$$F(\langle\hat{q}\rangle) = \frac{\mathrm{d}\langle\hat{p}\rangle}{\mathrm{d}t}. \tag{1.9}$$

When this "Ehrenfest substitution" is valid, this equation ensures that the centroid of the quantum wavepacket will follow the classical trajectory (where the classical trajectory is given, as usual, by Hamilton's equations of motion – the classical counterpart of Heisenberg's equations of motion given above).

Many textbooks on quantum mechanics give the impression that Ehrenfest's theorem adequately characterizes the classical limit. In fact, however, Equation (1.9) only holds under highly restricted circumstances. In order to use this characterization of the classical limit, one must be able to replace the mean values of functions, such as $\langle F(\hat{q})\rangle$, with a function of the means, $F(\langle\hat{q}\rangle)$. This substitution is legitimate only if the Hamiltonian of the system is a polynomial of second degree or less, such as in the case of linear or harmonic oscillator potentials. Such potentials, however, clearly do not cover all systems of physical interest.[25]

Leslie Ballentine has cogently argued that Ehrenfest's theorem is neither a necessary nor sufficient condition for characterizing the classical regime. To demonstrate that Ehrenfest's theorem is not sufficient for the classical limit, he reexamines the paradigm case of the harmonic oscillator. Although this is a system for which the "Ehrenfest substitution" is legitimate and the average values of the coordinates and momenta can be said to follow the classical equations of motion, "we know that a

[25] An example of a system for which the Ehrenfest's substitution is *not* valid is the case of a particle scattering off of a potential step; see Messiah (1965, p. 217) for a discussion.

quantum harmonic oscillator is not equivalent to a classical harmonic oscillator. In particular, the thermal equilibrium energy, and hence the specific heat, is different for classical and quantum oscillators" (Ballentine *et al.* 1994, p. 2854). Hence even when the conditions for Ehrenfest's theorem are satisfied, the classical harmonic oscillator is still empirically distinguishable from the quantum harmonic oscillator.

Similarly, Ballentine demonstrates that Ehrenfest's theorem is not a necessary condition for the classical limit by showing that there are several physical systems whose quantum states may behave classically, even when the Ehrenfest conditions do not apply. Ballentine notes that one of the difficulties with using Ehrenfest's theorem to characterize the classical regime is that it assumes that the classical limit of a wavefunction is a single trajectory rather than an ensemble of classical trajectories. He concludes that "the failure of the mean position in the quantum state to follow a classical orbit often merely reflects the fact that the centroid of a classical ensemble need not follow a classical orbit" (Ballentine *et al.* 1994, p. 2854). Ballentine's comments here raises an important issue that is often neglected in discussions of the classical limit: not only do we need to find an appropriate characterization of the classical limit, but we also need to determine what it is precisely that quantum mechanics is supposed to reduce to. Nickles, in his discussion of reductionism$_2$, notes this difficulty as well:

> It is frequently difficult to know what entities, states, or processes to identify with which. This problem becomes especially thorny when one attempts to reduce nonstatistical theories like CM or phenomenological thermodynamics to statistical theories like QT or statistical mechanics.
>
> *(Nickles 1973, p. 193)*

The view that the classical limit of quantum mechanics is classical statistical mechanics is associated with what is called the "statistical" or "ensemble" interpretation of quantum mechanics; this view can be traced back as far as Einstein (see, for example, [1949] 1970), and finds contemporary support in the work of figures such as Ballentine (1970).[26] This view is not without its detractors, however, and this debate highlights yet another subtlety in trying to derive classical behavior from quantum theory.

As this review of approaches to the classical limit shows, the reduction$_2$ of quantum mechanics to classical mechanics is much more subtle and problematic than the simple statements $\hbar \to 0$ or $n \to \infty$ might lead us to believe. A consensus seems to be emerging in the physics and philosophy of physics literature that the limit of some parameter (or combination of parameters) alone is insufficient for characterizing the classical limit, and that an adequate account of how classical behavior emerges out of quantum

[26] Einstein's views on the statistical interpretation (as expressed at the Fifth Solvay Congress in 1927) are discussed briefly again in Section 3.5.

systems needs to incorporate the phenomenon of decoherence.[27] Examining whether the decoherence program succeeds in providing a quantum-to-classical panacea is the subject of the next section.

1.5 Decoherence and the case of Hyperion

The decoherence approach emphasizes that realistic quantum systems are not isolated and closed, but rather are open systems interacting with their environment. In this context, the term 'environment' is used to collectively refer to things such as air molecules, light (photons), and even cosmic background radiation, which are continually impinging on macroscopic objects, becoming entangled with them, and hence altering their behavior.[28] As is well known, quantum systems exhibit interference effects and can exist in a superposition of two distinct states (the most troubling of which are the so-called Schrödinger cat states, which involve superpositions of macroscopically distinct states). These interference effects are made possible by the phase coherence between the different possible states of the system. Decoherence, then, can be defined as the loss of this phase coherence and consequent suppression of these interference effects through the interactions of a system with its environment.[29]

Although decoherence first became popular as a potential solution to the infamous measurement problem,[30] Wojciech Zurek and Juan Pablo Paz have more recently proffered decoherence as a solution to one of the central problems of quantum chaos: the potentially rapid divergence between the predictions of quantum mechanics and the (experimentally well-confirmed) predictions of classical mechanics when the classical system is chaotic. The time at which this divergence of predictions is predicted to become significant is generally referred to as the "break time," $t_\hbar$, which Zurek and Paz calculate via

$$t_\hbar = \lambda^{-1} \ln \frac{\Delta p_0 \chi}{\hbar}, \tag{1.10}$$

[27] See, for example, Zurek (2003), Berry (2001), and Landsman (2007). For an important opposing view see, for example, Ballentine (2004).

[28] More generally, "environment" in this context is used to describe any external or internal degrees of freedom with which the system of interest can become entangled. An example of internal degrees of freedom functioning as the environment would be the random motion of particles within the macroscopic object of interest (see, for example, Park and Kim (2003) and references therein).

[29] For a first introduction to decoherence see Bacciagaluppi (2004), and for a more technical extended introduction see Zurek (2003).

[30] The measurement problem, very briefly, is the following: the unitary evolution described in the Schrödinger equation predicts that a superposition of quantum states will lead to a superposition of macroscopic states for measuring apparatuses, cats, etc., and this is in conflict with the fact that we never observe such superpositions. For a discussion of the implications of decoherence for the measurement problem see Zurek (1991). For a criticism of the view that decoherence solves the measurement problem see Adler (2003).

where λ is the classical Lyapunov exponent, Δp_0 is a measure of the initial momentum spread of the Wigner function, and χ is a measure of the scale over which the potential, V, is nonlinear.[31] One can more easily estimate the break time by means of the analogous formula (Zurek 2003, p. 727 and references therein):

$$t_r = \lambda^{-1} \ln \frac{I}{\hbar}, \tag{1.11}$$

where I is the classical action. Although the action for a macroscopic object will typically be astronomically large (for example, on the order of 10^{77} (Zurek 2003, p. 727) or 10^{58} Planck units (Berry 2001, p. 46) for a moon), taking the logarithm of this number means that the break time can turn out to be surprisingly small. If the break time for a particular system is longer than, say, the age of the solar system, then it poses no threat; if, however, the break time is relatively short, then the possibility of a troubling conflict between quantum predictions and our classical observations arises.

To illustrate more precisely how classical chaos can lead to a divergence of quantum and classical predictions on a short timescale, Zurek (1998; 2003; Zurek and Paz 1997) considers the phase space formulation of quantum mechanics based on the Wigner function.[32] Recall that, although the Wigner function is a representation of the quantum state in phase space, it cannot be interpreted straightforwardly as a probability distribution because it can take on negative values. The evolution of the Wigner function $W(x, p)$, is given by the Moyal bracket as $\dot{W} = \{H, W\}_{\mathrm{MB}}$. When the potential, V, in the Hamiltonian, H, is analytic, the Moyal bracket is given by the classical Poisson bracket (which governs the Liouville flow of the classical distribution function in phase space) plus some quantum corrections:

$$\dot{W} = \{H, W\}_{\mathrm{MB}} = \{H, W\}_{\mathrm{PB}} + \sum_n \frac{\hbar^{2n}(-1)^n}{2^{2n}(2n+1)!} \partial_x^{(2n+1)} V \partial_p^{(2n+1)} W. \tag{1.12}$$

Since all of the quantum effects are contained in the second term, there is an agreement between the quantum and classical predictions as long as these $\hbar$-dependent quantum correction terms, known as Moyal terms, are negligible. The problem, however, is that when the classical system is chaotic, these quantum correction terms are no longer negligible. More specifically, as Zurek explains,

[31] The Lyapunov exponent (also called stability exponent) is a measure of the sensitivity to initial conditions (i.e., how quickly nearby trajectories are diverging). As we shall discuss in more detail below, the Wigner function is a phase space representation of a quantum state.

[32] There are in fact several different *equivalent* formulations of quantum theory, beyond the better-known Heisenberg and Schrödinger formulations. The Wigner function provides a useful phase-space formulation of quantum theory. A very brief introduction to nine of these different formulations of quantum theory – including the phase-space formulation – can be found in Styer *et al.* (2002).

Figure 1.1 Photograph of Saturn's moon Hyperion, taken by the Cassini Orbiter during a flyby on September 26, 2005. Courtesy of NASA/JPL-Caltech.

Correction terms above will be negligible when $W(x, p)$ is a reasonably smooth function of p, that is when the higher derivatives of W with respect to momentum are small. However, the Poisson bracket alone predicts that, in the chaotic system, they will increase exponentially quickly as a result of the "squeezing" of W in momentum … Hence after [the break time] $t_\hbar$ quantum "corrections" will become comparable to the first classical term on the right hand side.

(Zurek 1998, pp. 187–8)

It is these no-longer negligible Moyal terms that lead to the divergence between classical and quantum predictions.

As an illustration of how chaos can lead to a divergence of predictions on a disturbingly short time scale, Zurek and Paz (1997) examine the break time for Hyperion, one of Saturn's many moons. Hyperion was first discovered in 1848 and was observed at relatively close range by the Cassini Orbiter spacecraft in 2005 (see photograph in Figure 1.1). It is approximately three times the size of the state of Massachusetts and has roughly the shape of a potato.[33] Due to its irregular shape and the joint gravitational influence of Saturn and Titan (Saturn's largest moon), Hyperion tumbles chaotically in its orbit.[34] Rather than having a regular rotational period like Earth's 24 hours, both Hyperion's rate of rotation and axis of rotation are chaotic. This means that if one lived on Hyperion, one would not be able to predict where or when the sun would rise or set.

Using the above formula for the break time, Zurek and Paz (1997; see also Zurek 1998; Zurek 2003) have calculated that the predictions of quantum mechanics will begin to diverge from the empirically well-confirmed predictions of classical

[33] The dimensions of Hyperion are roughly 360 × 280 × 225 km (or 223 × 174 × 140 mi), which make it the largest known irregularly shaped natural satellite. Further information about Hyperion can be found, for example, at http://saturn.jpl.nasa.gov/science/moons/moonDetails.cfm?pageID=6, or at www.solarviews.com/eng/hyperion.htm.

[34] Wisdom *et al.* (1984) first predicted that motion of Hyperion would be chaotic, and this has since been confirmed by Klavetter (1989a,b).

mechanics after a mere 20 years. Berry (2001) has obtained a similar estimate of 37 years for the break time of Hyperion using a slightly different formula. Given that the planets in our solar system are around 4.5 billion years old, one would expect that Hyperion would have begun behaving quantum mechanically a long, long time ago. Zurek and Paz conclude,

> Given that t_r [the break time] is obviously orders of magnitude less than Hyperion's age one would expect the moon to be in a very non-classical superposition, behaving in a flagrantly quantum manner. In particular, after a time of this order the phase angle characterizing the orientation of Hyperion should become coherently spread over macroscopically distinguishable orientations – the wavefunction would be a coherent superposition over at least a radian. This is certainly not the case, Hyperion's state and its evolution seem perfectly classical.
>
> *(Zurek and Paz 1997, p. 370)*

On the assumption that quantum mechanics is a correct and universal theory, Hyperion's motions should in principle be governed by the laws of quantum mechanics. However, when it comes to chaotic systems such as Hyperion, these quantum laws make predictions that are inconsistent with our observations. Moreover, this rapid divergence of quantum and classical predictions is not unique to the case of Hyperion, but rather is, to varying degrees, a problem for all classically chaotic systems.

At this juncture there are several interpretive options open to us, some less desirable than others.[35] First, we might say that quantum mechanics has been falsified – at least when it comes to describing classically chaotic systems; this approach has been taken most famously by Joseph Ford (Ford and Mantica 1992), and calls for either a wholesale replacement of quantum theory, or a drastic limiting of its domain of application, resulting in some form of theoretical pluralism. Second, one might retain the view that quantum mechanics is a correct and universal theory (applying to celestial objects as well as microscopic ones), and argue that some important quantum phenomenon has been left out of the calculation that, when included, is able to bring the quantum predictions back in to alignment with our classical observations. On this view the limit of some parameter (or combination of parameters) alone is insufficient for characterizing the classical limit. This is the approach taken by advocates of decoherence such as Zurek. Finally, a third approach is to say that an additional quantum phenomenon such as decoherence is not in fact necessary, and that, with a particular interpretation and analysis, standard closed quantum mechanics is able to yield predictions indistinguishable from classical mechanics – even for classically chaotic systems. This is the approach taken by Ballentine and collaborators (Ballentine 2004; Wiebe and Ballentine 2005;

[35] I do not intend this list to be exhaustive.

Emerson 2001). While I will not discuss the first approach here, the second and third approaches will be briefly discussed in turn.

Decoherence has been described as the "essential ingredient that enables us to solve the apparent paradox caused by the lack of validity of the correspondence principle for classical chaotic systems" (Zurek and Paz 1997, p. 376).[36] On this view, macroscopic objects such as Hyperion are strictly quantum mechanical, but they are continually interacting with their environment, and hence the phase coherence necessary for quantum superpositions is irrevocably lost through this coupling to the environment. Even in outer space, there are things such as interplanetary dust and photons from the sun reflecting off of Hyperion, which are thought to be sufficient to effect a localization of the quantum state to a classical position. Returning to the equation of motion for the Wigner function given in Equation (1.12) above, simplified models show that decoherence acts to limit the growth of the gradients of the Wigner function, thereby ensuring that the quantum correction terms will be small, and consequently the evolution will be governed by the classical Poisson bracket. As a physical process, decoherence takes a finite amount of time to occur, and hence it is also important to show that decoherence will occur on short enough timescales. Although the decoherence rate will depend on the particular system of interest, current evidence suggests that decoherence can occur on an extremely short timescale, and hence any initial discrepancy between the classical and quantum predictions will be undetectable (Zurek 1998, pp. 191–2).

More recently, however, Ballentine and collaborators have reexamined the case of Hyperion, and challenged Zurek's claim that decoherence plays an essential role in the emergence of classical behavior – even when it comes to classically chaotic systems. The controversy centers on the significance of the break time, and how, precisely, a departure from classicality should be calculated.[37] Much of the disagreement can be traced back to the debate, mentioned above, concerning whether one should be interested in the "Ehrenfest break time," when a spreading wave-packet no longer mimics Newtonian mechanics, or what might be called the "Liouville break time," when the quantum dynamics no longer mimics the classical Liouville dynamics.[38] Ballentine's charge is that Zurek fails to adequately distinguish between the numerically distinct Ehrenfest and Liouville break times, and incorrectly uses the Ehrenfest break time as the relevant criterion for the failure of classicality.

[36] The phrase "correspondence principle" is being used here to mean the requirement that quantum and classical expectation values agree; once again, we shall see in Section 4.2 that this is not Bohr's correspondence principle.

[37] For an overview see, for example, Emerson (2001, Chapter 1).

[38] Recall that Liouville dynamics describes the evolution of classical probability distributions in phase space, when, for example, the precise initial conditions of the system might not be known.

The difficulty in determining whether the quantum predictions deviate from the empirically well-confirmed classical ones is that, for a massive object like Hyperion, one cannot solve the Schrödinger equation even by brute numerical integration. Hence, the approach taken by Wiebe and Ballentine instead is to cast the Schrödinger equation into a dimensionless form, and then solve it for a range of the dimensionless parameter involving $\hbar$. As emphasized earlier, the limit $\hbar \rightarrow 0$ is really just shorthand for the limit of some dimensionless parameter specific to the system of interest that involves $\hbar$. For the case of Hyperion, this dimensionless parameter is $\beta = \hbar T/I_3$, where I_3 is the largest moment of inertia of Hyperion, and T is its orbital period around Saturn. This approach yields a set of scaling relations, which Wiebe and Ballentine then use to draw inferences about the quantum predictions for Hyperion's behavior.

In their article, Wiebe and Ballentine consider two measures of the divergence between quantum and classical predictions. The first is the difference between the quantum and classical average angular momenta:

$$\left|\langle J_z\rangle_{\mathrm{Q}} - \langle J_z\rangle_{\mathrm{C}}\right|. \tag{1.13}$$

Their analysis shows that this obeys a $\beta^{2/3}$ scaling relation. Plugging in Hyperion's orbital period of $T = 1.8 \times 10^6$ seconds (about 21 days), a largest moment of inertia of $I_3 = 2.1 \times 10^{29}$ kg m^2, and $\hbar$ yields a value of $\beta = 9.3 \times 10^{-58}$ for Hyperion. Hence, the maximum quantum–classical difference for the dimensionless angular momentum of Hyperion should be around 5×10^{-37}, which is not an observable difference, and so not one that requires the invocation of decoherence to reconcile with our observations.

As an alternative, and perhaps more accurate, measure of the divergence of predictions, Wiebe and Ballentine consider the difference between the quantum and classical probability distributions given as:

$$|qm - cl|_1 = \sum_m \left|P_{\mathrm{cl}}(m) - P_{\mathrm{qm}}(m)\right|, \tag{1.14}$$

where the $P(m)$ are the angular momentum (m) distributions regarded as vectors.[39] They note that the quantum probability distributions do not converge pointwise to the classical probability distributions for either chaotic or non-chaotic states. Closer inspection, however, reveals that the quantum probability distribution is made up of a fine-scale oscillation superimposed on a smooth background, and this smooth background *does* converge to the classical distribution. Taking decoherence into

[39] The subscript "1" indicates that they are considering the 1-norm. See Wiebe and Ballentine (2005, p. 5) for further details.

account, they argue that the maximum difference between these probability distributions scales as

$$\max\left(|qm - cl|_1\right) \propto \left(\frac{\beta^2}{D}\right)^{1/6}, \tag{1.15}$$

where, for the case of Hyperion, D is the momentum diffusion parameter for the rotation of the moon due to collisions with the interplanetary dust around Saturn.[40] They conclude, "At the macroscopic scale of Hyperion, the primary effect of decoherence is to destroy a fine structure that is anyhow much finer than could ever be resolved by measurement" (Wiebe and Ballentine 2005, p. 13). In other words, the quantum and classical probability distributions are already observationally indistinguishable, and the effect of decoherence is only to bring them into even closer agreement by erasing this unobservable fine-scale structure. They conclude that "the quantum theory of the chaotic tumbling motion of Hyperion will agree with the classical theory, even without taking account of the effect of the environment ... [I]t is not correct to assert that environmental decoherence is the root cause of the appearance of the classical world" (Wiebe and Ballentine 2005, p. 13). More recently, however, Wiebe and Ballentine's conclusion has been challenged as inadequate by Schlosshauer (2006). Although I will not pursue this debate further here, what I think it does illustrate is that – even setting the measurement problem aside – the explanation of the emergence of the classical world out of quantum theory is far from a trivial or already accomplished task.[41]

1.6 Conclusion

Those who claim that classical mechanics simply reduces to quantum mechanics in the $\hbar \rightarrow 0$ limit are, at best, engaged in a drastic oversimplification. In discussing the relation between classical and quantum mechanics, the physicist Asher Peres recommends the following warning label: "Warning: The so-called principle of correspondence, which relates classical and quantum dynamics, is tricky and elusive" (Peres 1995, p. 229). More recently, Nicolaas Landsman, in his entry on the relation between classical and quantum mechanics, concludes, "For even if it is granted that decoherence yields the disappearance of superpositions of Schrödinger cat type ... this by no means suffices to explain the emergence of classical phase spaces and flows thereon determined by classical equations of motion ... A full

[40] The density of the interplanetary dust around Saturn was determined on the Voyager II mission to Saturn in 1981.
[41] While this debate may not, in the first instance, be about the measurement problem, one might argue that it is lurking not far below the surface. Indeed there are those who would argue that, when it comes to quantum mechanics, it is *all* about the measurement problem.

explanation of the classical world from quantum theory is still in its infancy" (Landsman 2007, p. 530). Of course the failure of current attempts to reduce classical mechanics to quantum mechanics does not mean that such a reduction is impossible. More importantly, perhaps, the many attempts to find such a reduction have been theoretically fruitful.

However, there are two important issues that I think have been obscured by this prevailing characterization of the relationship between classical and quantum mechanics in terms of reductionism$_2$ and decoherence. The first can be indicated by a play on the title of Zurek and Paz's 1997 article: "Why We Don't Need Quantum Planetary Dynamics." Their argument, recall, is that we don't need quantum mechanics to describe the motions of planets because decoherence is supposed to ensure that the predictions of quantum mechanics will be indistinguishable from the predictions of classical mechanics. One of the central theses of this book, however, is that we *do* need a quantum planetary dynamics, or rather what might better be described as a *planetary quantum dynamics*. In other words, while we might not need quantum dynamics to describe planetary systems, we do need planetary dynamics to describe quantum systems. Indeed much of the research in contemporary semiclassical mechanics can be described as the development of this planetary quantum dynamics. The defense of this thesis will be given in Chapters 5 and 6.

The second issue that I think has been obscured by a preoccupation with reductionism$_2$ and decoherence is that even if researchers succeed in reducing quantum mechanics to classical mechanics (with or without decoherence), this would still not be a complete account of the relationship between these two theories. Discussions of the classical limit and decoherence are specifically concerned with how to obtain a limiting agreement of predictions between the two theories. Although this is an important – and as we have seen nontrivial – question, it is a relatively narrow conception of the range of questions we might be interested in when we ask about the relationship between two theories. In the next three chapters I examine in detail the views of Werner Heisenberg, Paul Dirac and Niels Bohr on this very question of what is the relationship between classical and quantum mechanics. There we will see that these three figures defend a much richer variety of ways of thinking about intertheory relations than just the limiting agreement of predictions embodied in reductionism$_2$.

2

Heisenberg's closed theories and pluralistic realism

I could be bounded in a nutshell, and count myself a king of infinite space.
Shakespeare, Hamlet, *Act 2 Scene 2*

2.1 Introduction

As we saw in the last chapter, the received account of the relation between classical and quantum mechanics is a form of reductionism, where classical mechanics is supposed to emerge from quantum mechanics in the limit of some parameter.[1] When we examine more closely the views of three of the key founders of quantum theory, however, we see that none of them took the relation between these theories to be adequately captured by such a reductionist limit. Indeed as we shall see presently, Werner Heisenberg's account of the relation between classical and quantum mechanics is actually a strong form of theoretical pluralism, quite similar to Nancy Cartwright's metaphysical nomological pluralism introduced in Section 1.3.[2] Rather than viewing quantum mechanics as the fundamental theory that replaced classical mechanics, Heisenberg argues that both theories are required, each having its own proper domain of applicability and each being a perfectly accurate and final description of that domain.

Throughout his career, Heisenberg held a highly original view in the philosophy of science, centered on his notion of a closed theory. Very briefly, a closed theory is a tightly knit system of axioms, definitions, and laws that provides a perfectly accurate and final description of a certain limited domain of phenomena. This notion has profound implications for Heisenberg's understanding of scientific

[1] At least the predictions of classical mechanics are supposed to emerge out of the predictions of quantum mechanics, to within some reasonable degree of accuracy.

[2] As we shall see in Section 2.4, both Cartwright's and Heisenberg's pluralisms rest on the view that different domains of nature are governed by different systems of laws. Where they differ is in Heisenberg's claim that the theory describing one of these domains is a completely accurate and final description of that domain.

methodology, theory change, intertheoretic relations, and realism. Recognizing the central role that the concept of closed theories plays in Heisenberg's thinking reveals that he had a much more considered and consistent philosophy of science than has hitherto been assumed.[3]

Further insight into Heisenberg's philosophy of science can be gained through a comparison of his views with those of Thomas Kuhn. As Mara Beller briefly notes in her book *Quantum Dialogue*, there is a striking similarity between Heisenberg's account of closed theories and Kuhn's model of scientific revolutions (Beller 1999, p. 288). In Section 2.3, I explore this interesting parallel between Heisenberg and Kuhn in greater detail. My analysis draws on a little-known interview of Heisenberg, conducted by Kuhn in 1963 for the Archive for the History of Quantum Physics, in which Heisenberg mentions that he has just read Kuhn's recently published *Structure of Scientific Revolutions*. At this point, Kuhn breaks with his role as interviewer and begins to compare Heisenberg's views with his own. Their argument, in this interview, over whether Ptolemaic astronomy is to count as a closed theory helps to illustrate precisely where the similarities and dissimilarities between their views lie. By making use of this interview and Heisenberg's other writings, I show that while Heisenberg and Kuhn share a holistic account of theories, a revolutionary model of theory change, and even a notion of incommensurability, their views diverge fundamentally when it comes to the issue of scientific realism.

There is a widespread assumption that Werner Heisenberg is essentially a naive positivist or instrumentalist.[4] This reading of Heisenberg has two primary sources: First, his association with the so-called Copenhagen interpretation,[5] which is traditionally taken to view quantum theory simply as a "black box" tool for making predictions, and, second, the emphasis that Heisenberg himself places on observable quantities in his 1925 matrix mechanics paper and 1927 uncertainty paper.[6] As I shall argue in Section 2.4, however, this common reading of Heisenberg is mistaken. Not

[3] Important prior discussions of Heisenberg's closed theories include Chevalley (1988), Scheibe (1993), and Beller (1999). Beller (1999) in particular has emphasized the inconsistencies that appear to be in Heisenberg's philosophical thought.

[4] See for example Popper ([1963] 1989, p. 113).

[5] There is considerable confusion over what is meant by "the Copenhagen interpretation"; on the one hand it is used quite rightly to describe Bohr's own complementarity interpretation; on the other hand it has mistakenly been used to describe the standard physics textbook collapse interpretation. Nowhere in Bohr's writings does he ever appeal to the collapse of the wavefunction. For a discussion of how these two interpretations have been conflated see Howard (2004).

[6] In particular see Heisenberg ([1925] 1967, p. 261) and Heisenberg ([1927] 1983, p. 64). Olivier Darrigol has cogently argued that this emphasis on "observables only" did not in fact play a significant role in Heisenberg's discovery of quantum mechanics (Darrigol 1992, pp. 273–5). It is more likely that this emphasis was put in to please some of Heisenberg's colleagues, such as Wolfgang Pauli, who was the godson of the arch-positivist Ernst Mach. Although this latter connection is outside the scope of this book, Heisenberg's 1925 paper shall be discussed in more detail in Section 4.2.

only does Heisenberg explicitly reject positivism,[7] but as I shall show, he in fact endorses a version of scientific realism.

2.2 *Abgeschlossene Theorie*

Arguably the most central idea in Heisenberg's philosophy of science is his concept of a "closed theory." The German phrase that Heisenberg uses is "*abgeschlossene Theorie*," where "*abgeschlossene*" can be translated as "closed," "locked," "isolated," or "self-contained." Heisenberg defines a closed theory as a system of axioms, definitions, and laws whereby a field of phenomena can be described in a correct and noncontradictory fashion. Heisenberg first introduces the concept of a closed theory in connection with quantum theory as early as 1927: In the conclusion of his Solvay Conference paper with Max Born they write, "We wish to emphasize that ... we consider quantum mechanics to be a closed theory, whose fundamental physical and mathematical assumptions are no longer susceptible to any modification" (Born and Heisenberg [1927] 1928, p. 178).[8] From this point on, Heisenberg discusses closed theories regularly in his writings, up until his death in 1976. Throughout these almost fifty years his views on this topic seem to have changed very little.

Heisenberg identifies four theories of physics that have achieved the status of being closed: Newtonian mechanics, the nineteenth-century theory of heat (which he also calls statistical thermodynamics), Maxwell's electromagnetism together with optics and special relativity, and, of course, quantum theory. He notes that apart from these four areas, there remain other parts of physics, such as general relativity and elementary-particle physics, that are still "open."[9] Heisenberg justifies the choice of these four theories on the basis that "for each of these realms there is a precisely formulated system of concepts and axioms, whose propositions are strictly valid within the particular realm of experience they describe" (Heisenberg 1971, p. 98). Hence a necessary condition for a theory to be closed is that it have a precise axiomatic formulation.

Heisenberg had attended David Hilbert's lectures in Göttingen during the period 1922–24, and as Mélanie Frappier (2004) has shown in detail, his understanding of

[7] See for example Heisenberg's (1971) "Positivism, metaphysics and religion." There is, of course, the important question of whether Heisenberg changed his views, perhaps being a positivist in his younger days and only later becoming a realist. As I shall argue, however, the close connection between Heisenberg's account of closed theories and his pluralistic realism suggest that he was a realist at least as early as 1927, and maintained this position throughout the rest of life.

[8] I would like to thank Guido Bacciagaluppi for bringing this paper to my attention and for providing me with a copy of his English translation, from which this quotation is taken.

[9] Heisenberg offers the following justification for considering general relativity an open theory: "I think it is too early to count the general theory of relativity among the closed realms, because its axiom system is still unclear and its application to cosmological problems still admits of various solutions" (1971, p. 98).

closed theories was greatly influenced by Hilbert's axiomatic program. Heisenberg's recollections of this early influence from Hilbert on the development of quantum mechanics include the following:

> I remember that in being together with young mathematicians and listening to Hilbert's lectures and so on, I heard about the difficulties of the mathematicians. There it came up for the first time that one could have axioms for a logic that was different from classical logic and still was consistent. That I think was just the essential step.
>
> *(Heisenberg 1963, February 15th, p. 8)*[10]

By analogy, Heisenberg hoped to develop a consistent axiomatic system for quantum mechanics that was distinct from the consistent axiomatic system that describes classical mechanics.

Heisenberg describes the relation between the empirical world and the axiomatic system of a closed theory as follows: "It must be possible for the concepts initially stemming from experience to be so specified and fixed in their relations by definitions and axioms that mathematical symbols can be annexed to these concepts, among which symbols a consistent system of equations evolves" (Heisenberg [1948] 1974, p. 43). In other words, both axioms and definitions are required to properly link experience up with the abstract mathematical formalism of the theory. As Frappier notes, "Heisenberg's claim that the relations existing between physical concepts are defined in systems of axioms *and* definitions ... [might suggest] that he did not completely understand Hilbert's idea that a theory's axioms implicitly define its theoretical terms, and that consequently, a concept can change reference across interpretations" (Frappier 2004, p. 165). Frappier, however, offers an interesting alternative interpretation, according to which by "definitions" Heisenberg really means something closer to "interpretative principles."[11] Despite the importance of axiomatic systems for Heisenberg's account of closed theories, his account of such systems and its elements regrettably remains far from clear.

Heisenberg makes three striking claims about the nature of closed theories: first, that a closed theory covers a certain limited or bounded domain of phenomena; second, that it is a perfectly *accurate* description of that domain of phenomena; and third, that it is a *final* description – a permanent achievement that cannot be called into question by any future developments of science. For example, in his 1948 *Dialectica* article, he explains:

[10] These references with specific dates are to transcripts of oral history interviews for the Archive for the History of Quantum Physics. For further information about this archive and where the various duplicate deposits of this archive can be found, see www.amphilsoc.org/library/guides/ahqp/.

[11] For a fuller discussion of these issues I refer the reader to Frappier's excellent dissertation.

What, then, finally, is the truth content of a closed-off theory? ... The closed-off theory holds for all time; wherever experience can be described by means of the concepts of this theory, even in the most distant future, the laws of this theory will always prove correct.

(Heisenberg [1948] 1974, p. 45)

He reiterates these points again in the 1963 interview with Kuhn:

Newtonian mechanics is a limited description of nature and in that limited field it is perfectly accurate. It can never be improved. All attempts to improve Newtonian mechanics are just fruitless ... Since it is a closed axiomatic system I think it should be left as it is ... Of course it doesn't cover the whole of physics. There are other schemes. Already Maxwell theory is entirely different from it and again that is a closed scheme and cannot be improved.

(Heisenberg 1963, February 27th, pp. 21–2)

From the perspective of contemporary philosophy of science these three claims are surprising for the following reasons. First, theories such as quantum mechanics are typically thought of as being *universal* theories; that is, *in principle* the motions of the planets should also be governed by the laws of quantum mechanics, even though *in practice* such a description would be impractical, if not impossible. Indeed we saw this assumption at work in the discussion in Section 1.5, in which one of the fundamental challenges for decoherence is to explain why Saturn's moon Hyperion *looks* classical despite its really being quantum mechanical. Second, Heisenberg's claim that these theories are "perfectly accurate" flies in the face of the conventional wisdom that all theories involve idealizations, and are best thought of as providing models of the phenomena, rather than perfectly accurate descriptions. And third, his claim of "finality" is at odds with the received view that, for example, quantum mechanics is the fundamental mechanical theory that *replaced* classical mechanics; indeed someday it is likely that quantum mechanics will itself be replaced by a more fundamental theory of quantum gravity. On the standard view, there is no domain of phenomena in the world for which we can be certain that there will not someday be a more fundamental level of description.

Regarding his belief that closed theories are not universal, but rather only describe a limited domain, Heisenberg offers the following justification: the domain of application of a closed theory is determined by where the theory's concepts apply, and the process of axiomatizing the theory necessarily restricts the domain of applicability of the concepts. He describes the process by which axiomatization restricts the domain as follows:

For so long as concepts stem directly from experience ... they remain firmly linked to the phenomena and change along with them; they are compliant, as it were, toward nature. As soon as they are axiomatized, they become rigid and detach themselves from experience. To be sure, the system of concepts rendered precise by axioms is still very well adapted to a wide range of experiences; but we can never know in advance how far a concept established through definitions and relations will take us in our dealings with nature. Thus the axiomatization of concepts simultaneously sets a decisive limit to their field of application.

(Heisenberg [1948] 1974, p. 44)

The boundaries of a particular closed theory are not given a priori, but rather are discovered empirically. Heisenberg goes on to give the example of how some of the limits of the concepts of classical mechanics were found with the discovery of special relativity. Heisenberg's conclusion is that new developments in physics, such as the discovery of quantum mechanics, infringe not on the *validity* of classical mechanics, but only on its *scope*.[12]

Specifying the domain of applicability of the concepts of a closed theory is also important for his claim of accuracy. For example, Heisenberg frequently writes that wherever the concepts of Newtonian mechanics apply, there the laws of Newtonian mechanics will always prove to be correct (see, for example, Heisenberg [1935] 1979, pp. 41–2). Regrettably, however, he does not provide an independent way for determining where the concepts apply, other than the criterion that the laws of the theory accurately describe the phenomena. Heisenberg then combines the empirical accuracy of closed theories with the fact that they can be given an axiomatic formulation to justify his third claim of finality. He writes that the finality of closed theories is secured by "their compactness and manifold confirmation by experiment" (Heisenberg [1972] 1983, p. 124), where by "compactness" he means that a wide range of phenomena can be accounted for by means of a small number of postulates and axioms. This claim of finality leads him to conclude, for example, that "Newton's laws hold quite rigorously, and nothing in this will be changed for the next hundred thousand years" (Heisenberg 1971, p. 96; or for "millions of years from now and in the remotest star systems" see Heisenberg [1970] 1990, p. 186). In other words, Heisenberg believes that there is no possible future discovery that could ever call the laws of classical mechanics into question – at least no discovery in the domain of classical phenomena. The problem, once again, is that we seem to have no independent way of knowing what that domain is.

2.3 Holism, incommensurability, and revolutionary theory change

Heisenberg's closed theories exhibit a sort of holism – an interconnectedness of elements that resists small changes or modifications. In a 1958 celebration of the centenary of Max Planck's birth, Heisenberg explains the holistic nature of his closed theories by means of the following analogy:

[12] See also Heisenberg ([1935] 1979, p. 42). As we shall see in Section 2.4, Heisenberg's commitment to the limited domain of closed theories can also be seen as arising out of his metaphysical view that *nature itself* is divided into various regions of reality, each governed by its own systems of laws.

> One finds [in closed theories] structures so linked and entangled with each other that it is really impossible to make further changes at any point without calling all the connections into question … We are reminded here of the artistic ribbon decorations of an Arab mosque, in which so many symmetries are realized all at once that it would be impossible to alter a single leaf without crucially disturbing the connection of the whole.
>
> *(Heisenberg [1958b] 1974, p. 28)*

Holism, on this view, amounts to the claim that the structures and elements of a closed theory exhibit such a strong interdependence that one cannot modify a single element without destroying the whole.

For Heisenberg, the holistic nature of a closed theory arises out of the fact that it can be given an axiomatic formulation. He emphasizes this point in his interview with Kuhn:

> When you have a number of axioms as Newton had in the first pages of his famous book, *Principia Mathematica*, then the words are not only defined by the ordinary use of the language, but they are defined by their connections. That is, you cannot change one word without ruining the whole thing. Everything is bound up.
>
> *(Heisenberg 1963, February 27th, p. 24)*

Heisenberg's claim here is that, in an axiomatized theory, the meanings of the scientific terms are no longer definable in isolation, but rather are inextricably tied to the meanings of other scientific terms within that theory. So, for example, one cannot define the classical notion of force, without also invoking the concepts of mass and acceleration (or equivalently, the time rate of change of momentum). Hence any attempt to tinker with the Newtonian concept of force, will also lead to changes in these other concepts as well.

Heisenberg's commitment to holism leads him to adopt a revolutionary and anti-gradualist model of theory change – one in many respects similar to Kuhn's.[13] The holistic nature of closed theories means that when a genuinely new phenomenon is discovered, it cannot be accommodated by simply modifying an existing closed theory. Heisenberg describes this situation in a 1934 article as follows:

> The transition in science from previously investigated fields of experience to new ones will never consist simply of the application of already known laws to these new fields. On the contrary, a really new field of experience will always lead to the crystallization of a new system of scientific concepts and laws … The advance from the parts already completed to those newly discovered, or to be newly erected, demands each time an intellectual jump, which cannot be achieved through the simple development of already existing knowledge.
>
> *(Heisenberg [1934] 1979, p. 25)*

[13] Even before Beller's book (1999, p. 288), this similarity between Heisenberg's and Kuhn's models of theory change was noted by Erhard Scheibe (1988) and Friedel Weinert (1994, p. 305). None of these authors, however, seem to be aware of Kuhn's 1963 interview of Heisenberg, in which the actors themselves explore the similarities between their views.

In other words, new phenomena require the formation of a new closed theory. One can compare this claim to Kuhn's view that "the transition from a paradigm in crisis to a new one … is far from a cumulative process, one achieved by an articulation or extension of the old paradigm. Rather it is a reconstruction of the field from new fundamentals" (Kuhn [1962] 1996, pp. 84–5). In both cases the need for a wholesale revolutionary change is tied to the holistic nature of the predecessor theory, which cannot be modified or extended in such a way as to account for the new phenomena.

The fact that these two share a revolutionary and anti-gradualist model of theory change quickly becomes clear in the course of the 1963 interview. Here Heisenberg mentions that he has just read Kuhn's recently published *Structure of Scientific Revolutions*, and notes this parallel between Kuhn's views and his own:

> I have studied your book already and it gave me great pleasure to see the way you use the term paradigm … Yes, the necessity is to break away … One always, in such a situation, is forced to cut the branch on which one is sitting.
>
> *(Heisenberg 1963, February 27th, p. 22)*

At the conclusion of their discussion of theory change, Kuhn agrees with Heisenberg that, "with regard to evolutionary views about the development of science, you and I say very much the same thing" (Kuhn 1963, February 28th, p. 3).

Heisenberg explicitly connects up his revolutionary model of theory change with his belief that, in order to make progress in science, one must abandon the concepts of the predecessor theory. In the interview with Kuhn he goes on to criticize many of the other founders of quantum theory who "were not able to do the other step which would have been absolutely necessary to come further and that is, to throw away the old physics, that is throw away all the classical concepts and replace them with new ones" (Heisenberg 1963, February 28th, p. 13). He emphasizes this point again in 1968 when he writes:

> The decisive step is always a rather discontinuous step. You can never hope to go by small steps nearer and nearer to the real theory; at one point you are bound to jump, you must really leave the old concepts and try something new … in any case you can't keep the old concepts.
>
> *(Heisenberg 1968, p. 39; HCW C2, p. 431)*[14]

It is particularly interesting to note here Heisenberg's emphasis on the necessity in quantum theory of abandoning the concepts of classical mechanics. Although Heisenberg does not mention Niels Bohr by name here, in light of Bohr's doctrine

[14] Throughout, "HCW" refers to *Werner Heisenberg Collected Works* followed by the relevant volume number.

of the indispensability of classical concepts (which we shall discuss in Section 4.4), it is clear he is the target of these remarks.[15]

Heisenberg's view that each closed theory makes use of its own distinct closely-knit system of concepts leads him not only to an anti-gradualist model of theory change, but also to a position very close to Kuhnian incommensurability.[16] In the *Structure* Kuhn describes a central component of his incommensurability thesis as follows:

> Within the new paradigm, old terms, concepts, and experiments fall into new relationships with each other ... To make the transition [from Newton's] to Einstein's universe, the whole conceptual web whose strands are space, time, matter, force, and so on, had to be shifted and laid down again on nature whole.
>
> *(Kuhn [1962] 1996, p. 149)*

Four years before Kuhn, Heisenberg expressed a remarkably similar view:

> New phenomena that had been observed could only be understood by new concepts which were adapted to the new phenomena ... These new concepts again could be connected in a closed system ... This problem arose at once when the theory of special relativity had been discovered. The concepts of space and time belonged to both Newtonian mechanics and to the theory of relativity. But space and time in Newtonian mechanics were independent; in the theory of relativity they were connected ... From this one could conclude that the concepts of Newtonian mechanics could not be applied to events in which there occurred velocities comparable to the velocity of light.
>
> *(Heisenberg 1958a, pp. 97–8)*

Again in a later article, in which Heisenberg is discussing revolutions in the history of science, or what he prefers to call "changes of thought pattern," he writes, "Quite generally we may say that a change of thought pattern becomes apparent when words acquire meanings different from those they had before and when new questions are asked" (Heisenberg [1969] 1990, p. 156). Although Heisenberg never uses the term "incommensurability," one can identify in quotations such as these two key elements of the incommensurability thesis. First, although different theories may employ the same terms, the meanings of those terms can change fundamentally. Second, each theory has its own system of concepts, and these concepts are related to each other and the world in such a way that one cannot import them into the context of a different theory.[17]

[15] Heisenberg's commitment to closed theories also led him to reject other features of Bohr's philosophy, such as the duality at the heart of complementarity, as we shall see in Section 4.3. Important differences such as these call into further question the unity of the so-called "Copenhagen Interpretation."

[16] Scheibe (1988, p. 159) also recognizes this similarity between Heisenberg's views and Kuhnian incommensurability, though he does not explore this comparison in any depth.

[17] On this second point recall the quotation, given above, regarding the concepts of Newtonian mechanics not being applicable to contexts in which the velocities are comparable to the velocity of light.

Fortunately one need not speculate on how Kuhn would have reacted to Heisenberg's notion of incommensurability, for Heisenberg brings up this very issue in the course of the 1963 interview. When explaining to Kuhn his view of closed theories Heisenberg introduces the following example:

> As soon as you come to velocities, near the velocity of light, then it is not only so that Newtonian physics does not apply, but the point is that you even don't know what you mean by "velocity." Well, as you well know, you cannot add two velocities and so on, so just the word "velocity" loses its immediate meaning. That, I think is a very characteristic feature of what I mean by close[d] system; that is, when you have such a system and you get disagreement with the facts; then it means that you can't use the words anymore. You just don't know how to talk.
>
> *(Heisenberg 1963, February 27th, p. 24)*

Kuhn's immediate response at this point in the interview is to say: "O.K. That I entirely agree with. But I think you look too narrowly at the group of systems within which one can have that experience" (Kuhn 1963, February 27th, p. 24). Kuhn clearly recognizes in Heisenberg's account elements close to his own view, though he seems to criticize Heisenberg for not applying this incommensurability thesis more widely.

2.4 Theoretical pluralism and realism

A central part of the folklore surrounding Heisenberg is that he is a positivist or instrumentalist.[18] As I shall show below, however, a more careful reading of Heisenberg's writings reveals that this view is mistaken. Part of the difficulty in interpreting Heisenberg on this issue is that he does not explicitly make use of the realism–instrumentalism terminology; hence some care needs to be exercised in drawing out of his statements the correct implications for the realism debate. One explanation for the frequent misreadings of Heisenberg on this issue is that he often makes positivistic-sounding statements when he is discussing his scientific methodology, that is, his views concerning how one should go about constructing new scientific theories.[19] It is important, however, to distinguish Heisenberg's methodological views from his views regarding the status of the final product of that methodology. Moreover, as we shall see at the end of Section 4.2, significant doubts can be raised about whether Heisenberg's declared emphasis on "observables only" in 1925 played any real role in his development of quantum mechanics. In arguing that Heisenberg is in fact a realist, I do not mean to imply that he espoused some sort of

[18] Evidence for this folk interpretation can, for example, be found in Cassidy (1992), Popper (1963, p. 113) and Jammer (1974 p. 58).

[19] This is evident both in his 1925 matrix mechanics paper and in his 1927 uncertainty paper, where he is still in the process of constructing the new quantum theory.

hidden-variable interpretation – a realist view need not be committed to hidden-variables.[20] In calling Heisenberg a realist about quantum theory, what I mean is that he believed that quantum mechanics is an accurate, final, and true description of the way the world really is.

An important part of Heisenberg's defense of realism is the sharp distinction he draws between, on the one hand, phenomenological theories along with what he calls the "the pragmatic philosophy" and, on the other hand, closed theories and his own philosophy. Throughout his writings, Heisenberg is quite critical of phenomenological theories. He discusses the limited function of phenomenological theories as follows:

> [Phenomenological theories] can be extremely successful insofar as they can sometimes give the exact results and consequently agree extremely well with experiments. Still, at the same time they do not give any real information about the physical content of the phenomenon, about those things which really happen.
>
> *(Heisenberg 1968, p. 33; HCW C2, p. 425)*

Phenomenological theories, then, are understood as just calculational tools, theories that "save the phenomena" while failing to give us any true insight into nature.

In a short article on the role of phenomenological theories in physics, Heisenberg notes how the realist and the instrumentalist will evaluate phenomenological theories differently.[21] He somewhat misleadingly has what he calls the "Platonist" stand in for the part of the realist.[22]

> One could at this point note that the physicist or astronomer is going to unconsciously evaluate phenomenological theories quite differently depending on whether he has had his philosophical views formed by pragmatism, or by another way of thinking, like Plato's theory of ideas. For one who has grown up in pragmatism, a phenomenological theory will be valued more highly the more successes it can exhibit and the more exact predictions it can make. However, one who has from early on been seized by the persuasiveness of Platonic thought will, above all else, judge the phenomenological theories by whether and to what extent they can lead to an understanding of the true relations.
>
> *(Heisenberg 1966, p. 168; HCW C2, p. 386; my translation)*

In other words, Heisenberg's Platonist is not just concerned with empirical adequacy, but also with an understanding of the way things really are, that is, what he calls the "true relations."

In his recollections of a 1929 conversation with Barton Hoag, an experimental physicist from Chicago, Heisenberg elaborates on the differences between his view

[20] For a cogent defense of a similar distinction between being a realist and holding a particular classical view see McMullin (1984, pp. 12–13).

[21] I thank Mélanie Frappier for bringing this paper to my attention.

[22] This association of the Platonist with the scientific realist occurs in Heisenberg's other philosophical writings as well.

and that of what he calls "the pragmatic philosophy." Heisenberg has Hoag, as the spokesperson for the pragmatic view, describe the position as follows:

> Perhaps you make the mistake of treating the laws of nature as absolutes, and you are therefore surprised when they have to be changed. To my mind [that is, Hoag's mind], even the term "natural law" is a glorification or sanctification of what is basically nothing but a practical prescription for dealing with nature.
>
> *(Heisenberg [paraphrasing Hoag] 1971, p. 95)*

The pragmatic philosophy that Heisenberg has Hoag describe here is quite clearly an instrumentalist interpretation of scientific laws. Heisenberg responds that he finds this view entirely unsatisfactory and goes on to try to convince Hoag of his own alternative view of closed theories. It is here that he presents one of his clearest endorsements of realism:

> If, as we must always do as a first step in theoretical physics, we combine the results of experiments and formulae and arrive at a phenomenological description of the processes involved, we gain the impression that we have invented the formulae ourselves. If, however, we chance upon one of those very simple, wide relationships that must later be incorporated into the axiom system ... then we are quite suddenly brought face to face with a relationship that has always existed, and that was quite obviously not invented by us or by anyone else. Such relationships are probably the real content of our science.
>
> *(Heisenberg 1971, p. 99)*

We see once again here the sharp contrast that he wants to draw between phenomenological theories and closed theories. While Heisenberg is willing to adopt an instrumentalist attitude toward phenomenological theories, he believes that closed theories should be interpreted realistically.

It is important to note that Heisenberg takes neither quantum mechanics nor classical mechanics to be a phenomenological theory. That is, he adopts a realist interpretation of classical Newtonian mechanics right alongside a realist interpretation of quantum theory. In his book *Physics and Philosophy* he emphasizes that, on his view, "the system of definitions and axioms [of Newtonian mechanics] ... is considered as describing an eternal structure of nature" (Heisenberg 1958a, p. 93). Note his emphasis here on Newtonian mechanics being not just a description of an abstract theoretical structure, but rather being a description of *nature*. In light of the standard view that quantum mechanics is the theory that falsified and replaced classical mechanics, it is difficult to understand how Heisenberg could hold a realist interpretation of *both* these theories. In a 1935 article he tries to explain: "When one considers the basis of modern physics, one finds that it really does not infringe on the validity of classical physics ... It is not the validity but only the applicability of the classical laws which is restricted by modern physics" (Heisenberg [1935] 1979, p. 42). This passage suggests that the way in which he maintains a realist interpretation of both classical and quantum mechanics is by denying the universality (but not validity) of these theories' laws; this view fits well with his understanding of a closed theory as the accurate and final description of a certain limited domain of phenomena.

This interpretation of Heisenberg's view as a pluralistic realism finds further support in his 1942 unpublished manuscript *The Order of Reality.*[23] In this manuscript, he introduces the idea of various distinct regions of reality corresponding to each of his closed theories:

> By the expression "region of reality" we mean a collection of nomological connections. Such a collection [of laws] must on the one hand form a solid unity, because otherwise one could not legitimately speak of a "region," while on the other hand, it must be able to demarcate itself clearly from other collections [of laws] in order to render possible a division of reality.
>
> *(Heisenberg [1942] 1998, p. 273; my translation)*

Later he explicitly ties these various regions of reality to his closed theories: "The collection of nomological connections that we describe using classical mechanics is closed in itself. One means by this that the collection [of laws] can be explicated in a system of concepts and axioms that contain no contradictions and can be said to be in a certain sense complete" (Heisenberg [1942] 1998, p. 284; my translation). Heisenberg's concern here with various regions of reality, speaks once again to his realist – not instrumentalist – interpretation of closed theories.

The picture that starts to emerge from Heisenberg's views seems to be something like what Nancy Cartwright has called *metaphysical nomological pluralism*. As we saw in Section 1.3, she defines this as the view that nature is divided into various different domains, with each domain being governed by its own distinct system of laws. Like Heisenberg, Cartwright's concern is to deny the universality (or fundamentality) of scientific laws, not their accuracy within some limited domain. The similarity between her views and Heisenberg's becomes even more evident in her discussion of classical mechanics which we encountered in the previous chapter:

> [I]t is generally assumed that we have discovered that quantum mechanics is true … and hence classical mechanics is false … All evidence points to the conclusion that … Nature is not reductive and single minded … and is happily running both classical and quantum mechanics side-by-side.
>
> *(Cartwright 1995, p. 361)*

[23] There is an interesting, though as of yet unexplored, connection between Heisenberg's metaphysical view of closed theories and Wolfgang von Goethe's philosophy of science. Not only does Goethe figure prominently in Heisenberg ([1941a] 1979), (1961), and (1967), but one-third of this 1942 unpublished manuscript is titled "Goethe's Regions of Reality" ("*Die Goethe'schen Bereiche der Wirklichkeit*"). Heisenberg begins this section of the 1942 manuscript with a lengthy quotation from Goethe in which reality is divided into various distinct regions or levels. Though Heisenberg does not identify the source of this Goethe quotation, I am grateful to Gregor Schiemann (personal communication) for finding it in Goethe's *Sämtliche Werke* Volume 35, p. 403 (Stuttgart: Cotta 1895).

Compare this to Heisenberg's claim that,

> We no longer say "Newtonian mechanics is false and must be replaced by quantum mechanics, which is correct." Instead we adopt the formula "Classical mechanics is a consistent self-enclosed scientific theory. It is a strictly 'correct' description of nature wherever its concepts can be applied."
>
> *(Heisenberg [1948] 1974, p. 43)*

Neither Heisenberg nor Cartwright holds the simplistic view that there is a sharp quantum–classical border, at some particular length scale for example, where all the objects on one side fall under the domain of classical mechanics, and all the objects on other side fall under quantum mechanics. Instead they both carve up their dappled world in a more subtle and, for lack of a better term, theoretical way. For Cartwright the division happens as follows: "any situation that does not resemble a model of the theory will not be governed by [that theory's] laws" (Cartwright 1995, p. 359). For Heisenberg, as we've seen, the division is determined by where the concepts of the theory apply. By interpreting their pluralism realistically, as where particular laws of nature are operating, it is not clear how they would handle situations where the physical system in question "somewhat resembles" the models of the theory, or the concepts of theory "apply to some extent." Furthermore, given their view that one and the same object can be described using both classical and quantum mechanics, and their denial of a reductive relation between these two theories, there is the fundamental problem of how one can be sure that our theoretical accounts of such objects will not lead to outright contradictions. Although Cartwright is well aware of this problem of consistency, neither she nor Heisenberg offers a clear solution to this problem.[24]

2.5 The case of Ptolemaic astronomy

In the course of Heisenberg's 1963 interview with Kuhn, the four themes of holism, revolutionary changes, incommensurability, and realism come together. Recall (from the discussion at the end of Section 2.3) that while Kuhn endorses Heisenberg's notion of incommensurability, he criticizes him for not applying this notion more widely. After Heisenberg finishes describing his account of closed theories in the interview, Kuhn makes the following comment: "Now what strikes me is that there are many more systems to be found in the history of physics which would satisfy your criteria. You could have Ptolemaic astronomy …" (Kuhn 1963, February 27th, p. 23).[25] Heisenberg immediately interjects that Ptolemaic astronomy

[24] For Cartwright's discussions of the problem of consistency see (1995, p. 360) or (1999, p. 33).

[25] Very briefly, Ptolemaic astronomy is a geocentric model of our planetary system, due to Claudius Ptolemy of Alexandria (*c.* 100 to *c.* 170 C.E.), in which the earth is motionless at the center and the sun and all the planets

is merely a phenomenological theory that fails to meet his criteria for closedness. Kuhn then raises the very natural question of why Newtonian mechanics would not also count as a phenomenological theory. Heisenberg responds that Newtonian mechanics is importantly different because it is a closed axiomatic system – it exhibits a tight interconnected system of concepts and axioms, such that a single element cannot be changed without destroying the entire system. He continues:

> And just because of this an experiment which cannot be described by this axiomatic system means something more radical than for Ptolemaic astronomy. In Ptolemy's case, if the orbit didn't fit, he could add other epicycles. But if an experiment does not fit in Newtonian physics, you don't know what you mean by the words.
>
> *(Heisenberg 1963, February 27th, p. 24)*

Heisenberg then goes on to give the example of the change in the meaning of "velocity" from classical mechanics to relativity that was quoted earlier. The ground of Heisenberg's claim here is his commitment to the view that Newtonian mechanics admits of an axiomatic formulation characteristic of closed theories in a way that Ptolemaic astronomy does not. While Heisenberg is trying to distinguish Newton's theory from Ptolemy's by appealing to the holistic nature of Newton's theory and the concomitant idea of incommensurability, Kuhn, by contrast, wants to extend these notions of holism and incommensurability to the case of Ptolemaic astronomy as well. Kuhn replies,

> I am perfectly sure that this is what happened to people who had trouble with the Ptolemaic system and that when you are now looking back and … seeing it as a phenomenological theory, this is in large part simply because you are too far now away from it. Just as you say, the problem one runs into in Newtonian mechanics when the velocity gets too high is … that one doesn't know what velocity is anymore. I want to say with respect to a nice little phrase like "the stability of the earth" exactly that same sort of problem happens in the transition out of the Ptolemaic system.
>
> *(Kuhn 1963, February 27th, p. 24)*

For Kuhn, there is fundamentally no dividing line between phenomenological theories and theories that should be interpreted realistically. The central challenge of the *Structure* is to realism – that is, a challenge to the idea that through these changes in paradigm, scientists are getting closer to the truth. All theories, for Kuhn, are essentially just tools for puzzle solving.

Kuhn goes on in this interview to note another important difference between Heisenberg's views and his own:

move in circular orbits around the earth. Ptolemy is famous for introducing epicycles into this model in an attempt to reconcile it with observations. Ptolemaic astronomy was the dominant theory up until Copernicus's introduction (in 1543) of what is roughly our present heliocentric model of the solar system. For an annotated introduction to the first eight chapters of Ptolemy's *Almagest* where this model is introduced, including an explanation of epicycles, see, for example, Crowe (1990).

As you perhaps see in the book [*Structure*], I am not very happy about taking out a subgroup of these important discarded theories and saying, "Now these few are still with us." … In the long run quantum mechanics must be the substitute for Newton in the sense that Newton is the substitute for Ptolemy and Aristotle.

(Kuhn 1963, February 27th, p. 25)

In this quotation we see quite clearly how Heisenberg's theoretical pluralism is at odds with what we might call Kuhn's serial theoretical monogamy.

Kuhn concludes the second day of his interview with Heisenberg on this topic by saying:

Most of the things you've said I thoroughly agree with and I'm delighted. There is one area in which I think we're in quite essential disagreement … I find it very hard to talk about this permanent content of the closed scientific theory … So much of what you say are things that I think of myself as also wanting to emphasize in this book. But on this sort of point I would find that this undercuts the whole point of view for me.

(Kuhn 1963, February 28th, p. 3)

Thus, while Heisenberg and Kuhn somewhat surprisingly share a holistic account of theories, a revolutionary model of theory change, and a notion of incommensurability, they disagree fundamentally over the epistemological status of these theories and the issue of scientific realism.

The similarities between Heisenberg and Kuhn on these issues of holism, revolutionary change, and incommensurability naturally lead one to wonder whether Kuhn found in Heisenberg's writings a source for his own views. In this key 1963 interview that I have discussed in some detail, however, there is no evidence of a direct causal influence. Kuhn does not express a prior familiarity with Heisenberg's account of closed theories, nor does he take this opportunity to discuss the origin of his own views on these topics. Beller (1999) on the other hand has argued for an indirect influence of Heisenberg on Kuhn via the work of N.R. Hanson. My aim, however, has not been to pursue this thesis, nor to speculate on the sources or origins of Kuhn's views. Instead, I have tried to show that a careful comparison of Heisenberg's and Kuhn's views – facilitated by this remarkable interview in which they had the opportunity to sit down and discuss their views face to face – can bring us to a deeper understanding of Heisenberg's philosophy of science.

In particular, this comparison provides a foundation from which to correct several widespread and persistent misunderstandings about Heisenberg's views. First, Heisenberg is neither a naive positivist nor an instrumentalist. Instead, he espouses an interesting and novel form of pluralistic realism. Second, this deeper understanding of Heisenberg reveals a striking contrast between his views and Bohr's, calling into further question the unity of the so-called Copenhagen interpretation. Finally, these parallels between Heisenberg's and Kuhn's views show that there is a richness in Heisenberg's thinking about the nature of theories, realism, and theory change worthy of more careful consideration. This examination of Heisenberg's account of closed theories not only leads to a new understanding of Heisenberg's philosophy, but also to an appreciation of the potential relevance of his philosophy of science for contemporary debates.

2.6 The disunity of science

Heisenberg's philosophy of closed theories has important ramifications for his views on intertheory relations in general, and the relationship between classical and quantum mechanics in particular. He repeatedly maintains that quantum mechanics is neither a falsification of classical mechanics nor an improvement upon it. However, when it comes to saying precisely what is the proper way to understand the relationship between classical and quantum mechanics, Heisenberg is surprisingly silent. Although he sometimes talks about the virtues of the unity of science (e.g., [1941b] 1979) it appears that he is unable to connect this up with his account of closed theories. In the course of the 1963 interview with Kuhn, when he is faced with trying to give an account of the relation between these two closed theories, Heisenberg merely falls back upon Bohr's indispensability of classical concepts: "Newtonian mechanics is a kind of a priori for quantum theory. It is a priori in the sense that it is the language which enables us to say what we observe" (Heisenberg 1963, February 27th, p. 22). As we have seen, however, it is far from clear that this Bohrian line can be made consistent with Heisenberg's view that closed theories are "complete in themselves."[26] Moreover, on several occasions, Heisenberg explicitly rejects the continued use of classical concepts in quantum theory, arguing that they must be "thrown away" and "replaced with new ones," and that failing to do so will lead "physicists into assumptions that harbor contradictions … or to an impenetrable tangle of semi-empirical formulae" (Heisenberg [1972] 1983, p. 127). It would seem that on Heisenberg's account of closed theories, classical concepts would have no role to play – a priori or otherwise – in quantum theory.

As Frappier (2004) notes, Heisenberg rarely talks about mathematical limits such as the classical limit. To my knowledge Heisenberg discusses these limits in connection with his account of closed theories only once, and then only in a remarkably cursory way. Regarding the question of what relations there might be between the various closed theories, he writes,

[26] As Beller (1996) notes, "Heisenberg and Born often supported Bohr's position, although it was incompatible with their own. Such public support is one of the major sources of contradictions in Heisenberg's and Born's writings" (p. 184).

The relations between these four sets of concepts can be indicated in the following way: The first set [Newtonian mechanics] is contained in the third [electrodynamics, magnetism, special relativity and optics] as the limiting case where the velocity of light can be considered as infinitely big, and is contained in the fourth [quantum theory] as the limiting case where Planck's constant of action can be considered as infinitely small. The first and partly the third set belong to the fourth as a priori for the description of experiments. The second set [theory of heat] can be connected with any of the other three sets without difficulty and is especially important in its connection with the fourth.

(Heisenberg 1958a, p. 100)[27]

Apart from this perfunctory account, he offers no further elaboration on these non-trivial intertheory connections.

I want to suggest that the terseness of Heisenberg's account here, and more generally the near absence of his discussion of limiting relations elsewhere, can be understood as a result of the fact that he was not satisfied with these canonical accounts of intertheory relations. Without a cogent alternative, however, Heisenberg merely falls back on these accounts despite the fact that they do not sit comfortably with his own philosophy of closed theories. We get some sense of this and Heisenberg's struggle over how to interpret these limiting relations in a 1938 paper on the theory of elementary particles. He notes that there are two ways to interpret or "formulate" limits such as $\hbar \to 0$ and $c \to \infty$. According to the first formulation, "The previous theories continued to have standing as intuitive limiting cases in which the velocity of light can be regarded as very large and the action quantum as very small. The constants c and $\hbar$ rather designate the limits in whose proximity our intuitive concepts can no longer be used without misgivings" (Heisenberg [1938] 1994, p. 245). Heisenberg then goes on to note that there is a second way that these limits are often interpreted or formulated: "This state of affairs has often been expressed by the statement that the earlier theories emerged from relativity theory and from quantum theory through the limiting process $c \to \infty$ and $h \to 0$." He immediately goes on to explain why this "emergence" interpretation is unsatisfactory:

However, this [second] formulation is not quite unproblematical because it can be correct only if certain quantities are held constant during this limiting process (e.g., in the transition from quantum mechanics to classical mechanics, the masses and charges of the elementary particles) … Incidentally, the opposite limiting process $h \to \infty$ or $c \to 0$, while holding the above quantities constant, leads to meaningless results.

(Heisenberg [1938] 1994, p. 245)

[27] It is interesting to note that Heisenberg speaks of these as relations between *sets of concepts*, not between theories or the mathematical formalisms of the theories.

In contrast to these two interpretations of the classical limit, Heisenberg offers an account of what he would prefer the relation between classical and quantum mechanics to be: "In a definitive theory, however, these quantities would be determined from the few universal constants of physics, and a change of the magnitude of the universal constants could change nothing at all in the form of the physical laws" (Heisenberg [1938] 1994, p. 245). Heisenberg seems to dislike the idea that a new set of laws would emerge out of another set of laws simply by tinkering with the relative magnitude of constants such as $\hbar$. Instead, what he seems to desire is a new closed theory that would determine the values of these quantities directly and not involve any asymptotic relations at all. But in the absence of such a definitive closed theory, Heisenberg concedes that "[i]t therefore appears more correct to stay with the first formulation and designate $\hbar$ and c simply as the limits which are set to the application of intuitive concepts" (Heisenberg [1938] 1994, p. 245). In other words, in the absence of a better alternative, the Bohrian interpretation of the classical limit in terms of the applicability of classical concepts is "more correct" than the view that classical mechanics is actually recovered out of quantum theory as $\hbar \to 0$.

Heisenberg's 1934 comment regarding the issue of intertheoretic relations seems truer to his mark. Here Heisenberg writes, "The edifice of exact science can hardly be looked upon as a consistent and coherent unit in the naive way we had hoped" (Heisenberg [1934] 1979, p. 25). By committing himself to a picture of theories as largely isolated, closed axiomatic systems with incommensurable concepts, Heisenberg has left little room for a substantive theory of intertheoretic relations. Much like Cartwright, he is, in the end, left with a dappled world.

As we shall see in the next chapter, Paul Dirac's philosophy of science is almost point by point the antithesis of Heisenberg's. While Heisenberg takes classical and quantum mechanics to be closed, perfectly accurate, and final descriptions, for Dirac these theories are "open," approximate, and with no part immune to future revision. Where Heisenberg argues for wholesale change, Dirac seeks piecemeal modifications; where Heisenberg believes that quantum theory must make a sharp break with classical mechanics, Dirac believes that there is a deep structural continuity that should be exploited; and finally, while Heisenberg's views lead to a picture of science as disunified, Dirac's views offer a new approach to the unity of science.

Figure 3.1 Photograph of Paul Dirac (left) and Werner Heisenberg, Chicago, USA, 1929. [Courtesy of AIP Emilio Segrè Visual Archives]

3

Dirac's open theories and the reciprocal correspondence principle

... in apprehension, how like a god! The beauty of the world ...

Shakespeare, Hamlet, *Act 2 Scene 2*

3.1 Open theories

When it comes to the issues of reductionism, scientific methodology, and theory change, the views of Werner Heisenberg and Paul Dirac diverge in fundamental and interesting ways.[1] They revisited their disagreements over these philosophical issues many times throughout their careers, and their disagreements can be most succinctly described as a debate over whether physical theories are "open" or "closed." As we saw in the last chapter, Heisenberg's belief that classical and quantum mechanics are closed leads him to view these theories as perfectly accurate within their domains, inalterable, and correct for all time. Although Dirac never uses the term, his own views on classical and quantum mechanics can be fruitfully understood as a rival account of "open theories." Dirac argues that even the most well-established parts of quantum theory are open to future revision; indeed he takes no part of physics to be a permanent achievement, correct for all time.[2] Instead of viewing classical mechanics as a theory that had been replaced, he sees it as a theory that should continue to be developed, modified, and extended.

Unlike Heisenberg, who views physics as a set of consistent axiomatic systems, Dirac sees physics as a discipline much closer to engineering.[3] In a 1962 interview

[1] As we shall see in Chapter 4, both Heisenberg's and Dirac's views differ yet again from Niels Bohr's philosophy of classical and quantum mechanics. One of the recurring themes of this book is that there is no unified "Copenhagen philosophy," to which all three of these figures subscribe.

[2] For Dirac there is no Lakatosian "hard core" of either quantum or classical mechanics, which is immune to revision (cf. Lakatos 1970); for example, in 1936 he published a paper suggesting that one should give up on the principle of energy conservation for atomic processes, and (as we shall see in more detail below) in 1951 he published a letter to the editor of the top science journal *Nature* arguing that the notion of an aether should be reintroduced.

[3] Heisenberg explicitly rejects this idea of Dirac's that progress in physics is like progress in engineering (Heisenberg 1971, p. 97).

by Thomas Kuhn for the Archive for the History of Quantum Physics we see Dirac embracing the idea that progress in physics is like progress in engineering:

> I think that this engineering education has influenced me very much in making me learn to tolerate approximations ... I learned that even a theory based on approximations could be a beautiful theory. I rather got to the idea that everything in nature was only approximate, and that one had to be satisfied with approximations, and that science would develop through getting continually more and more accurate approximations, but would never attain complete exactness ... As a result of that, I haven't been much interested in questions of mathematical logic ... I feel that these things are just not important, that the study of nature through getting ever improving approximations is the profitable line of procedure.
>
> *(Dirac 1962, April 1st, p. 1)*

There are a number of key points that emerge from these brief comments by Dirac. First, that Dirac's oft-quoted insistence on "beautiful theories" should not be interpreted as an insistence on perfectly accurate or exact theories.[4] Second, we see Dirac advocating a gradualist model of theory change, whereby progress in physics takes place through continually developing ever more accurate approximations. Third, Dirac seems rather dismissive of the ideals of mathematical logic, which, as we saw at the beginning of the last chapter, had made such an impression on Heisenberg.

For Dirac, neither quantum mechanics nor even classical mechanics has reached its final form. Unlike Heisenberg, who took classical mechanics to be essentially unchanged since the opening pages of the *Principia* were written, Dirac sees classical mechanics as a theory that has been modified and improved: "Classical mechanics is essentially the mechanics of Newton. This mechanics was very much developed by Lagrange, Hamilton, and others, and more recently it has been modified in its underlying ideas through the appearance of Relativity" (Dirac 1951a, p. 10). While Heisenberg took classical mechanics and relativity to be two distinct closed theories, for Dirac they belong to the same theory.[5] Similarly Dirac criticizes the Copenhagen theorists for claiming that quantum theory had attained its final form. In a 1929 letter to Bohr he writes,

> I am afraid I do not completely agree with your views. Although I believe that quantum mechanics has its limitations and will ultimately be replaced by something better, ... I cannot see any reason for thinking that quantum mechanics has already reached the limit of its development. I think it will undergo a number of small changes.
>
> *(Dirac's letter to Bohr, Dec. 9th, 1929, in AHQP; quoted in Kragh 1990, p. 92)*[6]

[4] Dirac (1977) also emphasizes this view that even a theory based on approximations can be beautiful in his recollections at the Enrico Fermi Summer School. In Section 3.4 I shall return to a discussion of what Dirac means by "beauty."

[5] In addition to relativity, Dirac also includes classical electrodynamics in with classical mechanics. These are then contrasted with quantum mechanics (which includes quantum electrodynamics).

[6] "AHQP" refers to the Archive for the History of Quantum Physics.

In this quotation we not only see Dirac's commitment to the open character of quantum theory, but also his endorsement of a gradualist model of theory change.[7] Throughout his life Dirac maintained the view that classical and quantum mechanics are open theories, and hence devoted much of his scientific career to developing and extending these theories in new ways.

3.2 Structures, analogies, and the reciprocal correspondence principle

The foundation of Dirac's scientific methodology is his view that there is a deep structural continuity between classical and quantum mechanics. As Silvan Schweber has emphasized, Dirac sees quantum mechanics as an extension of classical mechanics, not a radical break with it (Schweber 1994, pp. 17–18). Indeed, Dirac even goes so far as to describe quantum mechanics as simply the generalization of classical mechanics to a noncommutative algebra.[8] For example, in 1926 Dirac writes that there is only "one basic assumption of the classical theory which is false, and that if this assumption were removed and replaced by something more general, the whole of atomic theory would follow quite naturally" (Dirac 1926a, p. 561). Later in that same year, he again reiterates this view and explains what this one assumption is: "The laws of classical mechanics must be generalised when applied to atomic systems, the generalisation being that the commutative law of multiplication, as applied to dynamical variables must be replaced" (Dirac 1926b, p. 281). In his later recollections, Dirac emphasizes how this approach differed from the other founders of quantum theory, such as Heisenberg: "If you look up these early papers you will see that there is quite a difference in our styles, because in my work the noncommutation was the dominant idea. With the Göttingen School, the dominant idea was the use of quantities closely connected with experimental results and the noncommutation appeared as secondary and derived" (Dirac 1971, p. 30). While it is an exaggeration to say that quantum mechanics just is classical mechanics with this one assumption changed, these quotations nonetheless underscore how deep and thoroughgoing Dirac believed the structural continuity between these two theories to be.

Dirac often speaks of this structural continuity as a formal analogy. For example, in 1939 he writes,

> [T]here is an extremely close formal analogy between quantum mechanics and the old [classical] mechanics. In fact, it is remarkable how adaptable the old mechanics is to the

[7] Dirac's gradualist model of theory change will be discussed in more detail in Section 3.3.

[8] Niels Bohr also viewed quantum mechanics as a sort of generalization of classical mechanics, though as we shall see in the next chapter (Section 4.3), he took it to be a generalization in a very different way from Dirac.

generalization of non-commutative algebra. All the elegant features of the old mechanics can be carried over to the new mechanics, where they reappear with an enhanced beauty.

(Dirac 1939, p. 124; DCW, p. 909)[9]

This important part of Dirac's philosophy can be referred to as the *structural continuity thesis*: quantum mechanics preserves a large part of the formal structure of classical mechanics despite the fact that it is non-commutative. Dirac sees this structural continuity as largely arising from the way in which quantum mechanics was constructed. He explains,

Quantum mechanics was built up on a foundation of analogy with the Hamiltonian theory of classical mechanics. This is because the classical notion of canonical coordinates and momenta was found to be one with a very simple quantum analogue, as a result of which the whole of the classical Hamiltonian theory, which is just a structure built up on this notation, could be taken over in all its details into quantum mechanics.

(Dirac [1933] 2005, p. 113)

What makes Dirac's philosophy particularly interesting is that he does not see the extent of this analogy between classical and quantum mechanics as being static or fixed for all time; rather, it is an analogy or structural continuity that can continue to be developed and extended in new ways.

In the 1933 paper just quoted from, Dirac gives a nice example of how the analogy between classical and quantum mechanics can be extended in new ways. He notes that in addition to the Hamiltonian formulation of classical mechanics, there is also a Lagrangian formulation, which is based on coordinates and velocities rather than coordinates and momenta. His aim in this paper is to figure out what the quantum analog to the Lagrangian formulation of classical mechanics would be, that is, to develop a *quantum* Lagrangian mechanics. Although the equations of Lagrangian mechanics cannot be taken over in a straightforward way, as in the case of Hamiltonian mechanics, he is nonetheless able to systematically set up a series of correspondence relations.

Dirac begins by showing that classical contact transformations (which transform a canonical set of coordinates and momenta into a new set of variables) are the analog of the quantum unitary transformations. In a later edition of his textbook Dirac explains,

The Heisenberg variables at time $t + \delta t$ are connected with their values at time t by an infinitesimal unitary transformation … The variation with time of the Heisenberg dynamical variables may thus be looked upon as the continuous unfolding of a unitary transformation. In classical mechanics the dynamical variables at time $t + \delta t$ are connected with their values at time t by an infinitesimal contact transformation and the whole motion may be looked

[9] "DCW" here refers to Dirac's Collected Works, edited by Dalitz, where many of Dirac's papers (those published prior to 1948) can be found.

upon as the continuous unfolding of a contact transformation. We have here the mathematical foundation of the analogy between the classical and quantum equations of motion, and can develop it to bring out the quantum analogue of all the main features of the classical dynamics.

(Dirac 1958, p. 125)

On the basis of this analogy between classical and quantum transformations, Dirac is able to show that the quantum "transformation function" $(q_t|q_T)$, which connects the two representations in which the dynamical variables q_t and q_T are diagonal, corresponds to e^{iS}, where S is equal to the time integral of the Lagrangian L, over a range of time from T to t.[10] In this 1933 paper, Dirac then writes down the following correspondence relations:

$$(q_t|q_T) \text{ corresponds to } \exp\left[\mathrm{i}\int_T^t L\mathrm{d}t/h\right] \equiv A(tT) \tag{3.1}$$

and hence for a very small (infinitesimal) time interval

$$(q_{t+dt}|q_t) \text{ corresponds to } \exp[L\mathrm{d}t/h]. \tag{3.2}$$

He concludes that "[t]he transformation functions in [3.1] and [3.2] are very fundamental things in quantum theory and it is satisfactory to find that they have their classical analogues, expressible simply in terms of the Lagrangian" (Dirac [1933] 2005, p. 117). By dividing the time interval up into a large number of very small time intervals, Dirac is then able to demonstrate the quantum analogue of the classical action principle, which says that Hamilton's principle function $S = \int_T^t L\mathrm{d}t$ remains stationary for small variations of the trajectory that keep the end points fixed. Indeed he then shows how the classical Lagrangian equations of motion are recovered from the corresponding quantum expression as $\hbar \to 0$.[11]

Historically, Dirac's development of this analogy between the Lagrangian formulation of classical mechanics and quantum theory was very important in laying the foundation for Richard Feynman's path integral formulation of quantum mechanics. In Feynman's 1965 Nobel prize lecture he recounts the now legendary story of how the physicist Herbet Jehle showed Feynman this 1933 paper of Dirac's, and Feynman, who was unable to understand what Dirac means by "corresponds" or "analogous," decided to set the quantum and classical expressions in Equation 3.1 equal to one another.[12] Thus, Dirac's pursuance and extension of the analogy

[10] Dirac's "transformation function" is essentially the Green's function; see Schulman (2005) for a brief discussion.

[11] Although this formulation of the classical limit is often attributed to Feynman, it first appears in this 1933 paper of Dirac's. For further details see Dirac ([1933] 2005, pp. 118–19), or Dirac (1958, pp. 128–9), or even Feynman (1948, p. 378).

[12] Although Feynman in 1965 retells the story as "equal," in his seminal 1948 paper he says "We now see that to sufficient approximations the two quantities may be taken to be proportional to each other" (Feynman 1948, p. 378). Dirac often seems to mean "proportional" when he says "analogous," such as in his discovery that the classical Poisson bracket is analogous to the quantum commutator (which shall be discussed below).

between classical and quantum mechanics turned out to be extremely fruitful, in that it led to Feynman's path integral formulation of quantum mechanics, and, as we shall see in Chapter 5, that in turn laid the foundation for the central methods of semiclassical mechanics.

For Dirac, the analogy between classical and quantum mechanics is important, not just for its own sake, but because it provides an important tool for the further development of physical theories. One can refer to this aspect of Dirac's philosophy of science as a *methodology of analogy extension*. In his well-known textbook *The Principles of Quantum Mechanics* Dirac describes this as his methodology quite explicitly:

> We should … expect to find that important concepts in classical mechanics correspond to important concepts in quantum mechanics, and, from an understanding of the general nature of the analogy between classical and quantum mechanics, we may hope to get laws and theories in quantum mechanics appearing as simple generalizations of well-known results in classical mechanics.
>
> *(Dirac 1958, p. 84)*

More concretely one can see this methodology at work in a 1945 paper called "On the analogy between classical and quantum mechanics."[13] Dirac begins by explaining that

> [t]here are two forms in which quantum mechanics can be expressed, based on Heisenberg's matrices and Schrödinger's wave functions respectively … The first is in close analogy with classical mechanics, as it may be obtained from classical mechanics simply by making the variables of classical mechanics into non-commuting quantities satisfying the correct commutation relations. The development of the analogy has been greatly hampered by the mathematical methods available for working with non-commuting quantities … The present paper will show how … in the case when the non-commuting quantities are observables, one can set up a theory of functions of them … [and] make closer the analogy between classical and quantum mechanics.
>
> *(Dirac 1945, p. 195)*

Dirac recognizes very well that extending the analogy between classical and quantum mechanics means not only developing new physics, but new mathematical methods as well. After Dirac has developed a general mathematical framework for dealing with functions of quantum observables, he then uses this new mathematics to ascribe a probability for non-commuting observables to have definite values, though he notes that this probability – being a complex number – cannot be given a straightforward physical interpretation. Perhaps surprisingly, he then goes on to show how this formal probability can be used "to set up a quantum picture rather close to the classical picture in which the coordinates q of a dynamical system have

[13] Feynman also cites this 1945 paper of Dirac as a forerunner of his path integral formulation of quantum theory.

definite values at any time" (Dirac 1945, p. 197). In other words, Dirac develops a way to reintroduce particle trajectories into quantum mechanics – something that he takes to be an important extension of the analogy between classical and quantum mechanics.

The importance that Dirac places on the analogy between classical and quantum mechanics can help us make sense of the dismissive attitude that he took towards Schrödinger's formulation of quantum theory. It is often thought that this critical stance towards Schrödinger was a product of some sort of allegiance on the part of Dirac to the Copenhagen–Göttingen School. Instead, I argue that Dirac's preference for the Heisenberg formulation is due to the fact that it makes explicit the structural similarities between the classical and quantum formalisms in a way that Schrödinger's formulation does not. Although Dirac is of course aware of the equivalence of these two formulations and notes the practical utility of Schrödinger's formulation, he maintains, "Heisenberg's form for the equations of motion is of value in providing an immediate analogy with classical mechanics and enabling us to see how various features of classical theory ... are translated into quantum theory" (Dirac 1958, pp. 115–16). Thus, while the different formulations of quantum mechanics might be mathematically equivalent, they are not equivalent when it comes to heuristics. It is this concern with heuristics – and not some sort of allegiance to Heisenberg or the Copenhagen interpretation – that leads him to favor Heisenberg's formulation.

As we have seen, Dirac takes there to be a strong formal or structural continuity between classical and quantum mechanics, which he thinks can be fruitfully extended in new directions. This methodology was referred to most broadly as one of analogy extension. However, the specific way that Dirac goes about extending this analogy can, in many physical contexts, be given a more precise characterization in terms of what I call the *reciprocal correspondence principle methodology.* This methodology can be understood as the interplay between the following two components: first, the general correspondence principle, which I take to mean quite broadly the use of classical mechanics in the further development of quantum theory,[14] and, second, what José Sánchez-Ron (1983) – in connection with the work of Adriaan Fokker – has called the "inverse correspondence principle," by which problems in quantum theory are used to guide the further development of

[14] As we shall see in the next chapter, there is considerable confusion over how exactly the correspondence principle should be understood. I have used the phrase "general correspondence principle" to try to distinguish it from Bohr's more specific (though multifaceted) correspondence principle, which, as I shall argue in Section 4.2, is a law-like relation between quantum transitions in an atom and particular harmonics in the Fourier expansion of the classical electron orbit. Both Bohr's specific correspondence principle and this general correspondence principle should again be distinguished from what Heinz Post (1971) calls the *generalized* correspondence principle, which in this context is the requirement that quantum mechanics "degenerates" into classical mechanics in the appropriate limit.

classical mechanics.[15] In my view, the general and inverse correspondence principles are inextricably intertwined in Dirac's work, and hence his methodology in these contexts is best described by the reciprocal correspondence principle. The reciprocal correspondence principle can be defined as the use of problems in quantum theory to guide the further development of classical mechanics *in order that* those results obtained in the classical context can then be transferred back into the quantum context to aid in the further development of quantum theory.[16]

In a rare moment of methodological reflection in a lecture at the 1949 Canadian Mathematical Congress, Dirac provides a clear account of what I have here called the reciprocal correspondence principle. The year in which Dirac is presenting this lecture suggests that he is concerned with the recent developments in quantum electrodynamics and quantum field theory. Dirac was notoriously unsatisfied with the divergencies or infinities plaguing quantum electrodynamics (QED), and with those who attempted to fix these problems with renormalization methods.[17] Although this is the context in which he is led to make his philosophy of scientific methodology explicit, I think it is nonetheless an accurate characterization of Dirac's methodology more generally. He explains,

> My own opinion is that we ought to search for a way of making fundamental changes not only in our present Quantum Mechanics, but actually in Classical Mechanics as well. Since Classical Mechanics and Quantum Mechanics are closely connected, I believe we may still learn from a further study of Classical Mechanics. In this point of view I differ from some theoretical physicists, in particular Bohr and Pauli.
>
> *(Dirac 1951a, p. 18)*

Although Dirac does not explicitly mention Heisenberg here, it is likely that he also has him in mind. In this quotation we see once again that Dirac not only takes quantum mechanics to be an open theory, capable of further modification, but that he takes classical mechanics to be open as well.

In order to see more clearly Dirac's expression of the reciprocal correspondence principle at work, and how his methodology differs from the Copenhagen–Göttingen group, it is worth quoting a bit further from this lecture. He begins, "It seems to me that one ought to separate the difficulties as far as possible … and attack each one separately." Notice the difference here from Heisenberg, who thought a hundred problems needed to be solved at once. Dirac continues,

[15] Although Sánchez-Ron (1983) is focusing on Fokker, he also notes that the inverse correspondence principle could be used to describe Dirac's methodology as well. Helge Kragh (1990), in his excellent biography of Dirac, also notes these similarities and applies Sánchez-Ron's inverse correspondence principle to Dirac.

[16] This is properly speaking a methodology not a principle, though in the case of both Dirac and Bohr, there are genuine correspondence *principles* underlying their methodologies that makes them possible.

[17] For a clear discussion of the history of Dirac's scientific views on QED and renormalization theory, see Kragh (1990).

For the purpose of this attack, we may very well work with the concepts of Classical Mechanics ... We want to find new ideas and I think it is very likely that some of the new ideas which we are looking for could be expressed and understood simply on the basis of Classical Mechanics ... In this respect I differ from Pauli, who believes that Classical Mechanics has nothing more to teach us in the Quantum Theory, and that to make advances one should work with the most highly developed form of our present Quantum Mechanics ... I believe that working with a highly developed form, which is necessarily a complicated form, restricts one's power for getting new ideas, and if there are new ideas which can be understood on a Classical basis, then one should try to work them out simply keeping to the Classical Theory, and after one has worked them out one can transfer them to the Quantum Theory by using the already established connection between Classical Mechanics and Quantum Mechanics.

(Dirac 1951a, p. 20)

In this last line we see clearly an important component of Dirac's reciprocal correspondence methodology, which can be referred to as the *thesis of transferability*. According to this thesis, any progress that we are able to make within the classical context will subsequently be transferable back into the quantum context. As we saw earlier, this two-way highway between classical and quantum mechanics is made possible by the thesis of structural continuity.

It should be emphasized that Dirac did not see these new required changes as being confined to the "frontiers" or "borderlands" of our theories. Rather, he thought they would reach to the very heartland. He exhorts, "The new ideas which I believe we should try to get may necessitate a fundamental change even in our description of those phenomena which can be quite well described by the existing quantum mechanics ... For this reason, I think that one should examine closely even the elementary and the satisfactory features of our present Quantum Mechanics" (Dirac 1951a, pp. 20–1). In other words, the open character of classical and quantum mechanics means for Dirac that there is no part of these theories that is a permanent achievement, correct for all time. As these quotations reveal, Dirac is willing to challenge the currently accepted parts of quantum theory in a way that Heisenberg, Pauli, and Bohr are not.

An example of Dirac's application of the reciprocal correspondence principle is found in his 1938 paper "Classical theory of radiating electrons."[18] In this work he writes,

If we accept Maxwell's theory, the field in the immediate neighborhood of the electron has an infinite mass. This difficulty has recently received much prominence in quantum mechanics ... where it appears as a divergence in the solution of the equations that describe

[18] I thank Matthias Frisch, who discusses the nonlocal character of the theory that Dirac develops here (Frisch 2002), for bringing this paper to my attention.

the interaction of an electron with an electromagnetic field ... One may think that this difficulty will be solved only by a better understanding of the structure of the electron according to quantum laws. However, it seems more reasonable to suppose that the electron is too simple a thing for the question of the laws governing its structure to arise, and thus quantum mechanics should not be needed for the solution of the difficulty. Some new physical idea is now required, an idea which should be intelligible both in the classical theory and in the quantum theory, and our easiest path of approach to it is to keep within the confines of the classical theory.

(Dirac 1938, p. 149)

In this quotation we can see the reciprocal correspondence principle at work. As a way of solving the problem of divergencies in *quantum* electrodynamics, Dirac turns to the development of a more adequate *classical* electrodynamics. Dirac's thesis of structural continuity – in this case between the classical and quantum theories of the electron – undergirds his thesis of transferability, which means that the classical solution to the problem of divergences that he obtains here will be transferable back to the quantum theory. In the above quotation he also offers, as a justification for his reciprocal correspondence methodology, the pragmatic point that the classical theory often provides a simpler context in which to work out new ideas.

3.3 A gradualist model of theory change

Dirac never abandoned the view that classical and quantum mechanics were open theories. Twenty-five years after his paper on the classical electron, he was still making use of the reciprocal correspondence principle. Although Dirac never accepted renormalization theory, he turned his attention to the problems that renormalization theory left unsolved. Among these he lists the following:

One of the problems is ... accounting for the number 137. Other problems are how to introduce the fundamental length to physics in some natural way, how to explain the ratios of the masses of the elementary particles and how to explain their other properties. I believe separate ideas will be needed to solve these distinct problems and that they will be solved one at a time through successive stages in the future evolution of physics. At this point I find myself in disagreement with most physicists. They are inclined to think one master idea will be discovered that will solve all these problems together.

(Dirac 1963, p. 50)

Clearly Heisenberg would be counted among those who believed these various problems needed to be solved all at once.

One of Dirac's more surprising approaches to solving these problems involved reintroducing an aether. Once again, he took the key to solving a quantum problem to lie in the development of a more adequate classical theory. In 1951 he had developed yet another classical electrodynamics, one that required postulating a velocity field defined at all points of space-time. Dirac interpreted this velocity as the velocity of the aether relative to the Earth. He argued that such an aether could be rendered consistent with relativity theory as long as one subjected the aether velocity to the quantum uncertainty relations (Dirac 1951b, p. 906). In this way Dirac was able to recover the Lorentz invariance of his theory. When, in 1952, Leopold Infeld pointed out that one could accept all of the conclusions of Dirac's new electrodynamics without postulating an aether, Dirac responded as follows: "Infeld has shown how the field equations of my new electrodynamics can be written so as not to require an aether. This is not sufficient to make a complete dynamical theory. It is necessary to set up an action principle and to get a Hamiltonian formulation of the equations suitable for quantization purposes, and for this the aether velocity is required" (Dirac 1952). For Dirac, the Poisson bracket correspondence that he had discovered in 1925 provided an important link between classical and quantum mechanics.[19] One can only take an advantage of this correspondence if one has a Hamiltonian version of the classical theory. Thus in his search for a new QED, his strategy was to develop an appropriate Hamiltonian version of classical electrodynamics, which could then be quantized. If this meant reintroducing an aether and absolute simultaneity, then he was willing to do this (see also Dirac 1953).[20] This reinforces the fact that, for Dirac, even the most accepted and well-established parts of theories were open to future revision.

When confronted with these same difficulties of QED, Heisenberg, by contrast, attempted to solve all of these problems at once by restricting himself to observables only – the same trick that had worked for him in 1925. This approach led Heisenberg to abandon quantum field theory in favor of the S-matrix program. For Dirac, on the other hand, agreement with experiments was not the final test of a theory. Regarding renormalization theory he writes, "Just because the results happen to be in agreement with observation does not prove that one's theory is correct" (Dirac 1987, p. 196). Dirac was quite critical of both renormalization and S-matrix theory. Despite the progress of the S-matrix school, he believed that high-energy physics should be based instead on equations of motion. As I shall show next, one can only make sense of Dirac's insistence on equations of motion in the context of his more fundamental belief in the unity of physics.

[19] The 1925 paper in which Dirac demonstrates the correspondence between the classical Poisson bracket and the quantum commutator shall be discussed in more detail below.

[20] Insofar as this book is about the relation between classical and quantum mechanics, I have been unable to fully explore Dirac's views on general relativity and cosmology. Silvan Schweber (personal communication) has rightly pointed out that Dirac's work on these space-time theories was likely an important influence on his more general philosophical views as well.

3.4 Beauty and the unity of science

In a 1970 paper Dirac links the following three key themes together with his quest for the unity of physics: the relative unimportance of agreement with experiment, the necessity of equations of motion, and mathematical beauty. He writes,

> If we believe in the unity of physics, we should believe that the same basic ideas universally apply to all fields of physics. Should we not then use the equations of motion in high-energy as well as low-energy physics? I say we should. A theory with mathematical beauty is more likely to be correct than an ugly one that fits some experimental data.
>
> *(Dirac 1970, p. 29)*

Here we see Dirac arguing that mathematical beauty is far more important than agreement with experiments. By now it should come as no surprise that Heisenberg is as dismissive of Dirac's criterion of beauty as Dirac is of Heisenberg's emphasis on agreement with observations. Regarding the features of a closed theory, such as its compactness, Heisenberg writes,"This might tempt one to conclude that perhaps its mathematical simplicity and beauty, and thus ultimately an aesthetic criterion, have exerted a governing influence on the persuasive power of the closed theory. But this influence should not be over-estimated" (Heisenberg [1972] 1983, p. 127). Heisenberg then goes on to list a number of features of classical and quantum mechanics that make them neither simple nor particularly beautiful.

There are, however, two points in the above quotation by Dirac that require further clarification: The first is the question of why Dirac places so much emphasis on equations of motion, and the second is the perennial question of what Dirac means by "beauty." Regarding the first question, the answer can be found a little further on in this same paper. He explains,

> High-energy physics forms only a small fraction of the whole of physics. The theories of most fields, such as solid-state physics, spectroscopy of atoms and molecules, and chemical physics, are based, fairly satisfactorily, on equations of motion. We believe in the unity of physics. The equations of motion that are so successful for most of physics cannot be simply discarded for one branch of physics. Although these equations may need modification, perhaps involving different kinds of variables, one would still expect to retain the basic structure.
>
> *(Dirac 1970, p. 30)*

Thus, Dirac's commitment to equations of motion can be understood in terms of his more fundamental belief in the unity of physics. He describes a unified physics as one in which the same "basic structures" reappear (suitably adapted) in all the different branches. A branch of physics formulated without equations of motion will, in his view, remain disconnected from the rest of physics. While such a disconnect might be expected on the view that each branch of physics is a closed theory that is complete-in-itself, for an advocate of open theories there remains the hope that these various branches can someday be unified.

The second point in need of clarification is determining what Dirac means by mathematical beauty.[21] In a 1939 paper, Dirac makes it clear that the theoretical virtue of beauty is distinct from the theoretical virtue of simplicity. He writes,

> The research worker, in his efforts to express the fundamental laws of Nature in mathematical form, should strive mainly for mathematical beauty. He should still take simplicity into consideration in a subordinate way to beauty. … It often happens that the requirements of simplicity and of beauty are the same, but where they clash the latter must take precedence.
> *(Dirac 1939, p.124; DCW, p. 909)*

Although Dirac is clear that considerations of beauty are of paramount importance, he is less clear about what beauty means. In this same article, he notes that beauty in science "is a quality which cannot be defined, any more than beauty in art can be defined, but which people who study mathematics usually have no difficulty in appreciating" (Dirac 1939, p. 123; DCW, p. 908). Unfortunately this is no more than a "I know it when I see it" definition of beauty.

While the notion of beauty is notoriously vague and multifaceted, I want to argue that one often-overlooked component of what Dirac means by beauty is a continuity and structural similarity with classical mechanics. Or, to put it more precisely, much of the beauty that quantum mechanics has is "inherited" from classical mechanics, and arises from the fact that a large part of the formal structure of classical mechanics is preserved in the transition to quantum theory. This aspect of his understanding of beauty is expressed most clearly in a 1927 unpublished manuscript where he remarks,

> The quantum theory has now reached a form … in which it is as beautiful, and in certain respects more beautiful than the classical theory. This has been brought about by the fact that the new quantum theory requires very few changes from the classical theory, these changes being of a fundamental nature, so that many features of the classical theory *to which it owes its attractiveness* can be taken over unchanged into the quantum theory
> *(Dirac 1927, emphasis added; quoted in Darrigol 1992, p. 345)*

For Dirac, classical mechanics is a beautiful theory that remains a benchmark against which all other physical theories are to be measured. This does not mean that he considered it to be the most beautiful scientific theory; indeed he soon came to believe that quantum mechanics was even more beautiful than classical mechanics (e.g., Dirac [1948] 1995, p. 1249). However when a theory departs unnecessarily from

[21] For a discussion of the role of beauty in science, including a discussion of Dirac's commitment to beauty, see McAllister (1996).

classical mechanics, then it loses some of its beauty; and this can be understood, in part, as a result of Dirac's fundamental commitment to the unity of science.

Although Dirac is deeply committed to the unity of science, he is not a reductionist in the usual sense of the term. More specifically, he does not think that the relation between classical and quantum mechanics is captured by an asymptotic agreement of predictions – what we described in Chapter 1 as Nickles' reductionism$_2$ relation. Instead, the all-important relation between these theories, for Dirac, is the fact that they share, to a great extent, the same formal structure. Remarkably, we see Dirac expressing this view of intertheory relations as early as 1925. Recall that in 1925 Dirac showed that the "Heisenberg product," (or what we now – following Dirac's terminology – call the commutator) that appears in Heisenberg's matrix mechanics paper is equal to $i\hbar$ times the classical Poisson bracket.[22] In modern notation this is expressed as

$$[\hat{x}, \hat{y}] = i\hbar\{x, y\}_{PB}. \tag{3.3}$$

In this paper Dirac explains, "the correspondence between the quantum and classical theories lies not so much in the limiting agreement when $h \rightarrow 0$ as in the fact that the mathematical operators in the two theories obey in many cases the same laws" (Dirac 1925, p. 649). In saying that the equations obey the same laws, Dirac is not simply pointing to some superficial similarity; rather, as Olivier Darrigol emphasizes, "The correspondence between two theories, he [Dirac] believed, was not limited to the form of the fundamental equations; it concerned mathematical *structures*, in the modern sense of the word" (Darrigol 1992, p. 317). Although this structural continuity might lead to an asymptotic agreement of predictions, to characterize the relation between these theories as consisting in this agreement is to miss Dirac's point that it is the continuity of structure that is fundamental. Indeed this structural continuity is one that holds even in a regime far removed from the classical limit. The structural continuities that Dirac identifies are ones that pervade these theories, not simply ones that emerge in some limit. In light of this, it seems that Dirac's view of the relation between classical and quantum mechanics is neither a form of reductionism nor pluralism, but, rather, what might be called a *structural account of intertheory relations*. On this view, the relation between two theories is

[22] Very loosely, the Poisson bracket can be thought of as the fundamental relation that generates the commutative algebra of classical mechanics. Formally the Poisson bracket is defined as

$$[F, G] = \sum_{\alpha=1}^{n} \left(\frac{\partial F}{\partial q_\alpha} \frac{\partial G}{\partial p_\alpha} - \frac{\partial F}{\partial p_\alpha} \frac{\partial G}{\partial q_\alpha} \right),$$

where F and G are functions of the generalized coordinates q and the generalized momenta p for a system with n degrees of freedom. See almost any text book on classical mechanics for further details.

characterized by the fact that they share a formal structure. I shall return to discuss this structural approach to intertheory relations in more detail in Chapter 7.

3.5 Dirac and the Einstein–Bohr debate

As we have seen in some detail, there is almost no overlap between Dirac's and Heisenberg's views on the philosophy of science. The question I want to examine next is whether there is any sense in which Dirac can be thought of as an adherent of Bohr's Copenhagen interpretation of quantum mechanics. Although Dirac carried out most of his research in England, the answer is typically assumed to be yes: Dirac's interpretive views on quantum mechanics – in so far as he had any – were squarely with Heisenberg and Bohr in the so-called Copenhagen–Göttingen camp. It is also typically assumed that Dirac was uninterested in the "Einstein–Bohr Debate," which was a long-standing philosophical debate over various aspects of quantum theory, consisting primarily of a series of challenges posed by Einstein, and Bohr's responses to these challenges. One finds this characterization of Dirac's views, for example, in Helge Kragh's biography of Dirac, in which he writes, "By and large, Dirac shared the positivist and instrumentalist attitude of the Copenhagen-Göttingen camp … Dirac never showed any interest in the opposition waged against Bohr's views by Einstein, Schrödinger, or de Broglie … Basically he was not very interested in the interpretational debate" (Kragh 1990, pp. 80–1). I shall argue that both of these assumptions about Dirac are mistaken: not only was Dirac interested in the Einstein–Bohr debate, but contrary to popular opinion, he sided with Einstein, against the Copenhagen–Göttingen camp.

At the center of my argument is an unpublished lecture of Dirac's entitled "Einstein and Bohr: The Great Controversy," in which Dirac makes explicit his allegiance to Einstein in this debate. After discussing this lecture in some depth, I shall show that this was not just a temporary lapse in Dirac's better judgment, but rather is a coherent part of his broader philosophical views concerning quantum theory as they developed over his career.

Before turning to Dirac's views on the Einstein–Bohr debate it is helpful to briefly review the central issues at stake in this debate. The debate between Einstein and Bohr began as a series of informal discussions in 1927 at the Fifth Solvay Congress. Our sources for this stage of the debate are the discussion in the proceedings of the Solvay meeting, Bohr's later recollections of this encounter in 1949, and a contemporaneous letter from Paul Ehrenfest who was present at the debates. In the discussions at the congress, Einstein proposes a simple experiment involving a screen with a small aperture through which electrons pass, and some photographic film in the shape of semicircle on the other side. He notes that there are two different possible interpretations one can take regarding the domain of validity of quantum mechanics. According to interpretation I, "the theory does not give any information about the individual processes, but only about an ensemble of an infinity of elementary processes" (Einstein [1927] 1928, p. 254; translated and reprinted in BCW 6, p. 101).[23] This is contrasted with interpretation II, according to which "the theory claims to be a complete theory of the individual processes." Einstein continues,

I have objections to make against interpretation II ... The interpretation according to which $|\psi|^2$ expresses the probability that *this* particle is situated at a certain place presupposes a very particular mechanism of action at a distance which would prevent the wave continuously distributed in space from acting at *two* places of the screen ... Interpretation II of $|\psi|^2$ in my opinion implies a contradiction with the relativity postulate.

(Einstein [1927] 1928, pp. 255–6; BCW 6, p. 102)

In these quotations we see that there is a cluster of concerns that Einstein is raising about the new quantum theory: first, whether quantum mechanics should be understood as a statistical theory about ensembles, or a theory of individual systems; second, whether quantum mechanics is a complete theory; and third, whether quantum mechanics involves an action at a distance that would conflict with relativity.[24] In Bohr's account of this discussion he also recalls Einstein "mockingly [asking] us whether we could really believe that the providential authorities took recourse to dice-playing" (Bohr 1949, p. 218) – a comment quite similar to the well-known one that Einstein had made to Max Born a year earlier (Einstein 1926). This comment suggests a fourth concern of Einstein's, namely that he objected to the fundamental indeterminism of the theory. Finally there is also a letter from Paul Ehrenfest, just two months after the congress, which gives further insight into the informal discussions that took place between Einstein and Bohr. Ehrenfest writes,

It was delightful for me to be present during the conversations between Bohr and Einstein. Like a game of chess. Einstein all the time with new examples. In a certain sense a sort of Perpetuum Mobile of the second kind to break the UNCERTAINTY RELATION. Bohr from out of the philosophical smoke clouds constantly searching for the tools to crush one example after the other. Einstein like a jack-in-the-box: jumping out fresh every morning.

(Letter from Ehrenfest to Goudsmit, Uhlenbeck and Dieke, November 3rd, 1927; BCW 6, p. 38; emphasis original)

This letter reveals a fifth concern, namely that, in these informal discussions, Einstein was challenging the fundamental status of the uncertainty relations.

[23] "BCW" stands for *Niels Bohr Collected Works*.

[24] Arthur Fine (1996), in his excellent book *The Shaky Game*, uses Einstein's unpublished 1927 critique of wave mechanics, to identify an even broader list of concerns that Einstein had regarding quantum theory in this year.

The next episode of the Einstein–Bohr debate took place three years later at the Sixth Solvay Congress. Regrettably this stage of the debate did not make it into the published proceedings of the congress, and we must once again rely on Bohr's recollection of the discussions and the correspondence of the participants after the congress. At this meeting Einstein introduced another thought experiment, which has come to be known as the "photon box" thought experiment. It involves a box with a shutter controlled by a clock that would emit a single photon at a precisely determinable time. According to Bohr's recollections, Einstein introduced this device as a way to beat the uncertainty principle. Using the relation $E = mc^2$, one can, after registering the time of the photon's emission, also precisely determine the energy of the photon by weighing the box before and after the emission. Such a simultaneously precise determination of the energy and time of the photon, would amount to a violation of the energy–time uncertainty principle. Bohr's response to this thought experiment involved showing that a consideration of the effect of the change in the gravitational field on the rate of the clock, reveals that such a method cannot in fact be used to beat the uncertainty principle.

There is, however, considerable controversy concerning whether Bohr correctly understood the aim of Einstein's photon-box thought experiment, and hence whether his account of this stage of the debate is accurate. Don Howard (1990) has cogently argued that the photon box thought experiment that Einstein introduces at the Sixth Solvay Congress was intended as a proto-EPR argument. The central piece of supporting evidence for this interpretation comes from a letter that Ehrenfest wrote to Bohr in 1931:

> He [Einstein] said to me that, for a very long time already, he absolutely no longer doubted the uncertainty relations, and that he thus, e.g., had BY NO MEANS invented the "weighable light-flash box" (let us call it L-F-box) "contra uncertainty relation," but for a totally different purpose.
>
> *(Ehrenfest 1931; quoted in Howard 1990, p. 98; emphasis Ehrenfest's)*

Bohr also alludes to this letter in his 1949 recollections, and suggests that Einstein had initially intended the photon box as a challenge to the uncertainty relations, but then showed that it could also be used to draw out another challenge to quantum theory. It is worth quoting Bohr at length:

> The Solvay meeting in 1930 was the last occasion where, in common discussions with Einstein, we could benefit from the stimulation and mediating influence of Ehrenfest, but shortly before his deeply deplored death in 1933 he told me that Einstein was far from satisfied and with his usual acuteness had discerned new aspects of the situation … In fact, by further examining the possibilities for the application of the balance arrangement, Einstein had perceived alternative procedures which, even if they did not allow the use he originally intended, might seem to enhance the paradoxes … After a preliminary weighing of the box with the clock and the subsequent escape of the photon, one was still left with the

choice of either repeating the weighing or opening the box and comparing the reading of the clock … Consequently, we are at this stage still free to choose whether we want to draw conclusions either about the energy of the photon or about the moment when it left the box. Without in any way interfering with the photon between its escape and its later interaction with other suitable measuring instruments, we are, thus, able to make accurate predictions pertaining *either* to the moment of its arrival *or* to the amount of energy liberated by its absorption. Since, however, according to the quantum-mechanical formalism, the specification of the state of an isolated particle cannot involve both a well-defined connection with the time scale and an accurate fixation of the energy, it might thus appear as if this formalism did not offer the means of an adequate description.

(Bohr 1949, pp. 228–9)

So certainly by 1949, Bohr had understood the photon-box thought experiment as a proto-EPR argument against the completeness of quantum theory, although he still – even after Ehrenfest's 1931 letter – believed that the original aim of this thought experiment was to undermine the uncertainty relations. It is unclear whether Bohr simply misunderstood Einstein on this point back in 1930, or whether Einstein had introduced the photon box in these discussions with more than one argument in mind. Either way, as Arthur Fine has argued, after 1930 – and perhaps even earlier if Howard is right – Einstein did finally accept the validity of the uncertainty relations, and the focus of the Einstein–Bohr debate turned to other aspects of quantum theory.[25]

The third – and most famous – chapter of the Einstein–Bohr debate is the well-known 1935 EPR paper, challenging the completeness of quantum theory, or as Einstein puts it in the 1949 Schilpp volume, forcing us to "relinquish one of the following two assertions: (1) the description by means of the ψ-function is *complete* (2) the real states of spatially separated objects are independent of each other" (Einstein 1949, p. 683).[26]

Even in this brief review of the Einstein–Bohr debate we can see that there is a number of interpretational issues being discussed, including completeness, determinism, the uncertainty relations, and nonlocality. After discussing Dirac's lecture on the Einstein–Bohr debate, I shall show that, not only did Dirac believe that quantum mechanics was incomplete and that a deterministic description of the microworld would be recovered, but even long after Einstein had accepted the validity of the uncertainty principle, Dirac argued that this principle would not survive in the future physics.

[25] Einstein alludes to his eventual acceptance of the uncertainty relations in the Schilpp volume: "… Heisenberg's indeterminacy-relation (the correctness of which is, from my own point of view, rightfully regarded as finally demonstrated) …" (Einstein 1949, p. 666).

[26] It is called the "EPR" paper after its three authors, Einstein, Podolsky, and Rosen. For Bohr's response to the EPR paper see Bohr (1935).

In the archives of the Paul A.M. Dirac Collection at Florida State University in Tallahassee, there is a handwritten lecture of Dirac's titled "Einstein and Bohr: The Great Controversy." It is written in a rough, choppy style and is apparently a lecture that was never delivered. The lecture is dated September 8, 1974, but unlike most of the other lectures there is no location given. The lecture begins,

> Einstein and Bohr had a controversy that lasted decades. Every time they met they would renew their argument. Neither was able to convince the other. Bohr had a school to support him. Einstein was alone, except for Schrödinger as a disciple.

Dirac's first characterization of the substance of the Einstein–Bohr debate is as a debate over determinism:

> Bohr as a protagonist for the standard quantum mechanics with a probabilistic interpretation. Einstein maintained that this could not be good theory. God does not play dice. To the end of his life he was seeking a unified field theory that would avoid probability.

Whether or not determinism was really what bothered Einstein about quantum theory, in Dirac's mind this was a central issue of the debate. Dirac then goes on to declare his allegiance to Einstein, and express his view that quantum mechanics is still not a final theory. He writes,

> My own belief was that Einstein was basically right, but he did not have a sufficiently general mathematical basis. The present quantum theory is an interim theory. It needs to be improved by a new theory of the future.

It is interesting that Dirac uses the past tense here, suggesting that this was also his view back in the early days of the debate. Not only does Dirac express his sympathies with Einstein's views, but he also takes the opportunity in this lecture to further distance himself from the Copenhagen interpretation:

> Bohr made the mistake (along with most physicists of his time) of being too complacent about the existing quantum mechanics. He accepted it as basically true and tried to build up a new philosophy – complementarity – to fit with it. Very ingenious and requiring profound thought – no one but Bohr could have done it. But still I do not believe nature works on those lines.[27]

This lecture suggests a very different picture of Dirac's philosophical views than the one that is commonly accepted. The Dirac of this lecture is neither a positivist uninterested in philosophical issues, nor an adherent of the orthodox Copenhagen interpretation. In what follows, I shall argue that the views expressed in this lecture are far more consistent with Dirac's other philosophical reflections on quantum theory than might be presupposed.

[27] Bohr's philosophy of quantum mechanics, as well as his viewpoint of complementarity, will be explained in the next chapter.

Dirac was present at both the 1927 and 1930 Solvay Congresses where the debate between Einstein and Bohr first took place. Since most of the debate was during informal discussions, there is, unfortunately, little in the published proceedings about the debate itself. Although it is typically assumed that Dirac was uninterested in this debate, his correspondence with Bohr just after the Sixth Solvay meeting reveals otherwise. In a letter to Bohr dated November 30, 1930, it becomes clear that not only was Dirac interested in the issues of the debate, but even more surprisingly, he was not yet fully convinced of Bohr's response to Einstein. He writes, "Dear Bohr, I would like to thank you for your interesting talks to me in Brussels about uncertainty relations … In looking over the question of the limit to the accuracy of determination of position, due to the limit c to the velocity of the shutter, I find I cannot get your result" (Dirac to Bohr, November 30, 1930; reproduced in Kragh 1990, pp. 85–6).[28] Although he was clearly interested in the debate at the time, it was not a topic that Dirac published on until much later in his career.

There are two primary published sources for Dirac's reflections on the Einstein–Bohr debate. The first is a 1963 *Scientific American* article titled "The Evolution of the Physicist's Picture of Nature." In this article Dirac identifies determinism and questions about the validity of the uncertainty relations as the two issues at the heart of the Einstein–Bohr debate. He writes,

> The hostility some people have to the giving up of the deterministic picture can be centered on a much discussed paper by Einstein, Podolsky and Rosen dealing with the difficulty one has in forming a consistent picture … Everyone is agreed on the formalism. It works so well that nobody can afford to disagree with it. But still the picture that we are to set up behind this formalism is a subject of controversy. I should like to suggest that one not worry too much about this controversy. I feel very strongly that the stage physics has reached at the present day is not the final stage … The present stage of physical theory is merely a steppingstone.
>
> *(Dirac 1963, pp. 47–8)*

It is interesting to note that Dirac's characterization of the central challenge of the EPR paper is quite different from Einstein's. In Dirac's discussions of the debate, he seems to overlook the issue of nonlocality, which was arguably from the beginning the feature of quantum mechanics that most troubled Einstein. Instead, Dirac focuses on the issues of determinism and the uncertainty relations. It is not clear whether Dirac simply misunderstands Einstein on this point, or whether nonlocality was just not a feature of quantum theory that troubled him. Either way, it is important to emphasize that, as becomes clear in the above passage, Dirac's dismissal of the project of finding a consistent picture behind quantum theory is *not*

[28] Note that Dirac's recollection, just one month after the Sixth Solvay Congress, is that the informal discussions did include concerns over the validity of the uncertainty relations, though Einstein is not explicitly named.

based on some sort of positivism or instrumentalism, as has commonly been assumed. Rather, it is based on his belief that quantum mechanics is not a final theory. This is not an incompleteness in the EPR sense, but rather, as we saw in Section 3.1, part of Dirac's view that quantum mechanics is an open theory.

In this same 1963 article, Dirac uses the Einstein–Bohr debate as a platform from which to launch his own challenge to the fundamental status of the uncertainty relations. His argument is based not on a thought experiment to beat the uncertainty principle, as Einstein's was, but rather on a theoretical challenge to Planck's constant being a fundamental constant of nature.[29] Dirac's argument begins with the fine structure constant, which is a dimensionless number, very nearly equal to 1/137, involving Planck's constant $\hbar$, the speed of light c, and the electron charge e. He believes that the fine structure constant requires an explanation, and argues that not all of the constants $\hbar$, c, and e can be fundamental; rather, only two can be fundamental, and the third must be derived from the other two. After offering arguments for why the velocity of light and the electron charge must both be fundamental, he concludes,

> If $\hbar$ is a derived quantity instead of a fundamental one, our whole set of ideas about uncertainty will be altered: $\hbar$ is the fundamental quantity that occurs in the Heisenberg uncertainty relation connecting the amount of uncertainty in a position and a momentum. This uncertainty relation cannot play a fundamental role in a theory in which $\hbar$ itself is not a fundamental quantity. I think one can make a safe guess that uncertainty relations in their present form will not survive in the physics of the future.
>
> *(Dirac 1963, p. 49)*

Long after Einstein had accepted the uncertainty relations as fundamental, we see Dirac using the Einstein–Bohr debate as a platform from which to challenge the status of these relations, and argue that the present formulation of quantum theory will need to be superceded.

The second published source for Dirac's reflections on the Einstein–Bohr debate is an article that Dirac wrote for the Centennial Symposium on Einstein that was held in Jerusalem in 1979. In this article, titled "The early years of relativity," Dirac reflects back on the fifth Solvay Congress. He writes,

> I first met Einstein at the 1927 Solvay Conference. This was the beginning of the big discussion between Niels Bohr and Einstein, which centered on the interpretation of quantum mechanics. Bohr, backed up by a good many other physicists, insisted that one can use only a statistical interpretation, getting probabilities from the theory and then

[29] This was a shift in Dirac's views brought about by his investigations of relativity and cosmology; the early Dirac (as far as I know) did not challenge the fundamental status of Planck's constant. Schweber (personal communication) has suggested that Dirac's questioning of the fundamental status of Planck's constant contributed significantly to his understanding of quantum theory as "open." I am grateful to him for drawing out this important connection between fundamental constants and the status of a theory as open or closed.

comparing these probabilities with observation. Einstein insisted that nature does not work in this way, that there should be some underlying determinism.

(Dirac [1979] 1982, p. 84)

Dirac then goes on to describe the pressure one feels as a student to conform to Bohr's interpretation of quantum theory, adding that "once he has passed his exams, he may think more freely about it, and then he may be inclined to feel the force of Einstein's argument" (Dirac [1979] 1982, p. 84).

In this article Dirac not only argues that quantum mechanics is an interim theory that needs to be replaced, but that the theory that will replace quantum mechanics will most likely be a deterministic one.

It seems clear that the present quantum mechanics is not in its final form. Some further changes will be needed, just about as drastic as the changes made in passing from Bohr's orbit theory to quantum mechanics. Some day a new quantum mechanics, a relativistic one, will be discovered, in which we will not have these infinities occurring at all. It might very well be that the new quantum mechanics will have determinism in the way that Einstein wanted. This determinism will be introduced only at the expense of abandoning some other preconceptions that physicists now hold. So under these conditions I think it is very likely, or at any rate quite possible, that in the long run Einstein will turn out to be correct.

(Dirac [1979] 1982, p. 84)

From the beginning, Dirac was concerned with getting a relativistic quantum mechanics, and when he refers to quantum mechanics here, he means it quite broadly to include quantum field theories. In particular, it becomes clear in this passage that one of the issues troubling Dirac is the infinities plaguing quantum field theories. As we saw earlier, Dirac was an outspoken opponent of the renormalization procedures for handling infinities, and viewed these infinities as yet another indication that present quantum theories need to be replaced by a more adequate theory.

These writings of Dirac reveal that his unpublished lecture, "Einstein and Bohr: The Great Controversy," – rather than being anomalous – is in fact a consistent part of his general philosophical views concerning quantum theory. In particular, we see the following three key features of Dirac's philosophy of quantum mechanics echoed: First, quantum mechanics is not a complete theory, rather only an interim theory. Second, the theory that will replace quantum mechanics will most likely be a deterministic theory. Third, the uncertainty relations are not fundamental limitations, and they will not likely survive in the future physics.

One can perhaps understand the timing of Dirac's public writings on the Einstein–Bohr debate as a result of the fact that towards the end of his career, Dirac increasingly found himself in a position similar to the later Einstein; that is, largely isolated in going against the grain of the physics community in challenging the orthodox quantum theories. Although both Dirac and Einstein objected to the indeterminism inherent in quantum theory, and at different stages – and for quite different reasons – challenged the fundamental status of the uncertainty relations, the issues that seemed to trouble them the most were quite different; for Einstein it was arguably the nonlocality of standard quantum theory, whereas for Dirac it was the infinities occurring in quantum field theories. Nonetheless, both men can be counted in the same camp in this debate, in so far as they were both led to reject the received quantum theories, in their search for a new unified physical theory.

Both Dirac's published and unpublished reflections on the Einstein–Bohr debate reveal that he is neither a positivist uninterested in interpretational issues, nor an adherent to the Copenhagen–Göttingen camp, as is commonly assumed. In these writings it becomes clear that his disavowals of philosophy are not based on some sort of positivism, but rather on his view that quantum mechanics is not yet the correct theory. This, along with his explicit rejection of complementarity, suggests that there was far less philosophical agreement among the founders of quantum theory than is typically supposed.

Although Dirac's interpretation of quantum theory is quite different from Bohr's, in many respects his scientific methodology is much closer to Bohr's than it is to Heisenberg's. Both Dirac and Bohr sought to systematically utilize classical mechanics in the development of quantum theory, that is, to transcribe the results of classical mechanics into quantum mechanics in a way that is consistent with the fundamental contrast between these theories. The chief methodological difference between Dirac and Bohr consists in the extent to which each was willing to use classical mechanics. As we shall see in the next chapter, Bohr's use of classical mechanics is largely confined to the correspondence principle. Furthermore, by 1927 Bohr seemed to think that the new quantum theory had reached its final form, and hence classical mechanics was no longer required for its development.[30] Dirac, by contrast, sees a much wider role for classical mechanics in the development of quantum theory, as he notes in the following passage:

> The steady development of the quantum theory that has taken place … was made possible only by continual reference to the Correspondence Principle of Bohr … A masterful advance was made by Heisenberg in 1925, who showed how equations of classical physics could be taken over in a formal way … thereby establishing the Correspondence Principle on a quantitative basis and laying the foundations of the new Quantum Mechanics … This does not exhaust the sphere of usefulness of the classical theory.
>
> *(Dirac 1932, p. 453; DCW, p. 623)*

[30] As I shall argue in the next chapter, Bohr did think classical mechanics was still required after 1927 for the proper *interpretation* of quantum mechanics (e.g., he thought quantum mechanics depended conceptually on classical mechanics), but he did not think classical mechanics was required for its further development.

Not only did Dirac strive to use classical mechanics in a much more comprehensive way (as we saw, for example, in his development of a quantum Lagrangian mechanics), but his commitment to the open status of quantum theory meant that classical mechanics could be used in the further development of quantum mechanics long after 1927. In order to see more clearly how Dirac's use of classical mechanics differs from Bohr's, however, we must first take a closer look at what precisely the content of the correspondence principle is, and doing this is the central subject of the next chapter.

4

Bohr's generalization of classical mechanics

Something is rotten in the state of Denmark.

Shakespeare, Hamlet, *Act 1 Scene 4*

4.1 The rise and fall of the old quantum theory

The principles of the old quantum theory were first articulated by Niels Bohr in 1913 in a three-part paper titled "On the Constitution of Atoms and Molecules."[1] Bohr had adopted Ernest Rutherford's model of the atom, according to which the mass of the atom is concentrated in a small central nucleus, while the electrons orbit the nucleus in planetary trajectories. The key challenge facing Rutherford's model was that it was unstable: according to classical electrodynamics, the electron, which is a moving charged body, should radiate energy and so eventually collapse into the nucleus. Bohr's solution was to incorporate Max Planck's theory of radiation, according to which "the energy radiation from an atomic system does not take place in the continuous way assumed in ordinary electrodynamics, but that it, on the contrary, takes place in distinctly separated emissions" (Bohr 1913, p. 4; BCW 2, p. 164).[2] In this early paper, and most subsequent accounts, Bohr summarized his quantum theory by means of two assumptions or postulates. According to the first postulate, electrons cannot travel in any path they like around the nucleus, rather atomic systems can only exist in a series of discrete "stationary states," in which the electron is in a particular allowed stable periodic orbit and is not emitting radiation.

[1] This review is not intended to be a comprehensive history of the old quantum theory; rather my goal is two-fold: first to provide just enough of a scientific context to make sense of Bohr's philosophical views on the correspondence principle and the rational generalization thesis, which are the focus of this chapter, and second, to show how, in many respects, the old quantum theory can be seen as a "pre-history" of the problem of quantum chaos and modern semiclassical mechanics, which will be discussed in Chapter 5. For the interested reader, there are several excellent, more detailed histories of the old quantum theory, such as Jammer (1966), Darrigol (1992), and Mehra and Rechenberg (1982).

[2] Throughout "BCW" refers to *Niels Bohr Collected Works*, where most of Bohr's papers can be readily found, followed by the relevant volume number.

Intuitively, these stationary states can be thought of as a series of concentric circles or orbits around the nucleus, along which the electron travels; these stationary states are labeled by means of the principal quantum number n, with the lowest allowed orbit (the "normal" or ground state) labeled $n = 1$, the next stationary state of higher energy $n = 2$, and so on. Bohr notes that when the electron is in one of these stationary states, its motions can be adequately described by means of classical electrodynamics; however, when the electron makes a transition from one stationary state to another, classical electrodynamics no longer applies. The second postulate of Bohr's old quantum theory is that, when there is a transition between different stationary states, n' and n'', the emitted radiation (the photon) is of a single frequency, ν, that is given by the difference in the energy of the two states ($E_{n'} - E_{n''}$) divided by Planck's constant.

$$\nu = (E_{n'} - E_{n''})/h \tag{4.1}$$

This formula is typically referred to as the "Bohr–Einstein frequency condition." The second postulate constituted a significant break from the classical electrodynamics, according to which a variety of radiation frequencies would be emitted and those frequencies would be determined solely by the motion of the source.[3]

Bohr's two postulates are not yet enough to pick out from all of the classically allowed orbits those orbits corresponding to stationary states. In order to determine the stationary states the following "quantum condition" also needs to be introduced:

$$\oint p_\theta \mathrm{d}\theta = nh, \tag{4.2}$$

where the integral is taken over one period of the electron orbit, and p is the angular momentum (radial component), θ is the angle in the plane of the electron orbit, and n is the quantum number – a non-negative integer which picks out each of the stationary states.[4] As Max Jammer summarizes, applying the old quantum theory consists of essentially three steps: "first, the application of classical mechanics for the determination of the possible motions of the system; second, the imposition of certain quantum conditions for the selection of the actual or allowed motions; and third, the treatment of the radiative processes as transitions between allowed motions subject to the Bohr frequency formula" (Jammer 1966, p. 90). As Bohr was keenly aware, the old quantum theory was a thorough mixing of quantum and classical ideas, and one of Bohr's central goals was to show that, despite this

[3] These classical expectations for the radiation of an atom will be reviewed more precisely in the following section on the correspondence principle.

[4] This familiar way of expressing the quantum condition does not appear in Bohr (1913). For an excellent historical review of Bohr's work on the old quantum theory, including a discussion of Bohr's quantum conditions, see Darrigol (1992), Chapters V and VI.

blending of quantum and classical concepts, the resulting theory was nonetheless internally consistent. As I will show in the next section, it was the correspondence principle that led Bohr to believe that one could "anticipate an inner consistency for this [quantum] theory of a kind similar to the formal consistency of the classical theory" (Bohr 1924, p. 25; BCW 3, p. 482). Before turning to these interpretive problems, however, it will be useful to examine briefly the extent to which the old quantum theory could be generalized, and the limitations it encountered that ultimately led to its demise.

This initial formulation of the old quantum theory could only countenance the idealization of circular electron orbits, having only a single degree of freedom, and hence a single quantum condition corresponding to the quantization of angular momentum. The subsequent development of the old quantum theory was concerned with extending Bohr's theory to include, for example, more general types of electron orbits such as ellipses, which require two degrees of freedom: the angle θ and the radius r. In 1915 and 1916 Arnold Sommerfeld was able to generalize Bohr's quantum condition by proposing that the stationary states of a periodic system with N degrees of freedom are determined by the condition that "the phase integral for every coordinate is an integral multiple of the quantum of action" (Sommerfeld 1916, p. 9):[5]

$$\oint p_i \mathrm{d}q_i = n_i h \qquad \text{where } i = 1, 2, 3, \ldots N. \tag{4.3}$$

While Bohr's initial quantum condition was quite successful at explaining the main features of the hydrogen spectrum, it could not account for the so-called fine structure – the fact that on closer resolution the main spectral lines are themselves composed of a multiplet of lines. Sommerfeld was able to account for the fine structure of hydrogen by incorporating relativistic considerations (the variation of the electron mass in its elliptical orbit) into these generalized quantum conditions. By doing so, the old quantum theory could be applied not only to simply periodic motion, but to so-called conditionally (or multiply) periodic motion as well.[6]

Despite the great success of Sommerfeld's extension of Bohr's old quantum theory, there remained the problem of the choice of coordinates to be used in the quantum conditions: different choices of coordinates lead to different quantized

[5] Jammer notes in his consideration of this passage that not only was this extension of the quantum condition to multiple degrees of freedom independently proposed in the same year by both Wilson and Ishiwara, but by Planck indirectly as well, suggesting that it should perhaps be called the Sommerfeld–Wilson–Ishiwara–Planck quantum conditions (see Jammer 1966, p. 91 for references).

[6] An example of conditionally (or multiply) periodic motion taken from astronomy is the relativistic advance of the perihelion of Mercury, where, instead of having a simple ellipse that closes on itself, the perihelion (closest turning point of the motion) precesses, tracing out a rosette pattern.

Figure 4.1 Photograph of Arnold Sommerfeld (left) and Niels Bohr, Lund, Sweden, 1919. [Courtesy of AIP Emilio Segrè Visual Archives, Margrethe Bohr Collection]

orbits.[7] A solution to this difficulty was proposed in 1916 independently by Paul Epstein, a student of Sommerfeld's, and by the astronomer Karl Schwarzschild. Using techniques from celestial mechanics, Schwarzschild argued that the so-called action–angle variables were the appropriate coordinates for determining the stationary states of atoms.[8] In order to construct these variables, one needs to find a canonical transformation from the usual momentum–position coordinates (p, q) to the action–angle coordinates (J, θ). This is accomplished by finding a generating (or action) function S, such that the following relations hold:

$$p = \frac{\partial S(J, q)}{\partial q} \qquad \theta = \frac{\partial S(J, q)}{\partial J}. \tag{4.4}$$

In these coordinates, the action variables J_i are functions of the constants of the motion, and thus are the "first integrals" of the system; conjugate to these action variables are the angle variables, θ_i. Both Schwarzschild and Epstein proposed that the coordinates should be chosen such that the action function S, is separable, which is to say that S can be expressed as a sum over N action functions, each of which depends on only one coordinate. While the introduction of these techniques allowed the quantization conditions to be extended to more general conditionally periodic systems, as well as to account for the spectrum of a hydrogen atom in an

[7] For a more detailed history of these quantum conditions see Jammer (1966), Chapter 3.

[8] For a helpful introduction to the use of action–angle variables in the old quantum theory see Darrigol (1992), Chapter VI.

electric field (the Stark effect), they could only be applied to systems that are separable.[9] Unfortunately, however, most systems of physical interest are nonseparable.

The next development in the history of the quantum conditions, though now familiar to physicists working in quantum chaos, remains largely unknown to historians and philosophers of physics.[10] In 1917 Albert Einstein wrote a remarkable paper titled "On the quantum theorem of Sommerfeld and Epstein" in which he writes, "Notwithstanding the great successes that have been achieved by the Sommerfeld-Epstein extension of the quantum theorem for systems of several degrees of freedom, it still remains unsatisfying that one has to depend on the separation of variables" (Einstein 1917, p. 84; CPAE Vol. 6, p. 436).[11] Einstein notes that while the quantity $p dq$ is invariant (i.e., independent of the choice of coordinates) for one degree of freedom, as in Bohr's original quantum condition, when it comes to multiple degrees of freedom, the individual products $p_i \, dq_i$ in the Sommerfeld–Epstein–Schwarzschild quantum conditions are not invariant. Einstein proposes that the quantum conditions should instead be based on the sum of those products, $\sum p_i dq_i$, which is an invariant quantity, resulting in the following new quantum conditions:

$$\oint \sum_i p_i dq_i = n_i h \qquad \text{where } i = 1, 2, 3, \ldots N. \tag{4.5}$$

Here N denotes the number of degrees of freedom of the system, and the integral is taken over each of the topologically independent loops in the "rationalized coordinate space" Einstein introduces in this paper.[12]

Although this formulation of the quantum conditions by Einstein provided a solution to one of the central problems of the old quantum theory, what makes his paper truly remarkable is his subsequent comment. By the time Einstein received the proofs of this article he realized that his generalized quantum conditions are also inadequate. In a "Supplement in Proof" added to this paper Einstein writes,

> If there exist fewer than N integrals of type $[R_k(q_i, p_i) = \text{const}]$, as is the case, for example, according to Poincaré in the three-body problem, then the p_i are not expressible by the q_i and

[9] In a separable system there is no coupling between different degrees of freedom, and the Hamilton–Jacobi equation and action function can be written as a sum of terms each depending on only one pair of variables.

[10] For example, neither Jammer (1966) nor Darrigol (1992) mentions this paper of Einstein's. A notable exception is Batterman (1991).

[11] These page numbers refer to the English translation of *The Collected Papers of Albert Einstein*, abbreviated "CPAE."

[12] This multiply connected "rationalized coordinate space" is introduced in Section 5 of Einstein's paper, and is arguably a more important contribution than his extension of the quantum conditions. Today, this space is typically thought of as a torus in phase space, on which these quantization conditions are defined (see, for example, Tabor (1989), Chapter 6). For a further explication of Einstein's quantum conditions and this rationalized coordinate space, see Bokulich (2001) or Stone (2005).

the quantum condition of Sommerfeld-Epstein fails also in the slightly generalized form that has been given here.

(Einstein 1917, p. 92; CPAE Vol. 6, p. 443)[13]

If there are fewer constants of the motion ("integrals of type $R_k(q_i, p_i) = \text{const}$") than there are degrees of freedom, then a system is said to be nonintegrable and can exhibit chaotic behavior. What Einstein had recognized here is that for chaotic systems, such as Poincaré's infamous three-body problem, the quantum conditions of the old quantum theory break down. Although Einstein's quantum conditions received moderate attention, his warnings about the challenge that nonintegrable systems pose for the old quantum theory seem to have been largely ignored. This is perhaps not surprising since it is was only towards the end of the twentieth century that people began to realize how pervasive chaotic systems are in nature. What would ultimately lead to the downfall of the old quantum theory was not Einstein's theoretical arguments about nonintegrable systems in general, but rather the calculations of the helium atom – a particular nonintegrable system that would soon become the second-most famous three-body problem.[14]

Although the old quantum theory was remarkably successful in accounting for the hydrogen atom and a variety of other phenomena, it failed miserably when it came to the helium atom. In 1922, John van Vleck calculated the ionization potential of the ground ("normal") state of the helium atom and obtained a value that differed significantly from the experimental value. In a short paper summarizing his results he writes,

The study of helium, the simplest atom except hydrogen, should be a key to a generalized Bohr theory of atomic structure. However, no satisfactory model of normal helium has yet been devised, for the models … all give the wrong ionization potentials if the non-radiating orbits are determined by the Sommerfeld quantum conditions.

(van Vleck 1922, p. 419)

The models that van Vleck refers to involved proposing different particular configurations for the two-electron periodic orbits in the ground state of helium. Not only did van Vleck's calculations for the ground state of helium disagree with the experimental value, but a year later Hendrik Kramers published a strikingly similar incorrect value that he had obtained using a different, even more rigorous method.[15] These repeated failures to obtain the correct ground-state energy of the helium atom

[13] I have changed Einstein's "*l* integrals" to "*N* integrals" for consistency of notation.

[14] The most famous three-body problem is, of course, the Earth–Sun–Moon system, which has challenged physicists and astronomers since the time of Newton.

[15] For a more detailed account of this historical episode see Darrigol (1992) Chapter VIII or Mehra and Rechenberg (1982) Vol. 1 Part 2, Section IV.2.

began raising questions about the adequacy of the assumptions and quantum conditions of the old quantum theory.

Upon hearing of van Vleck's result, Heisenberg in 1922 wrote to Sommerfeld describing an approach to solving the helium atom, that involved assuming that the quantum number in the quantum conditions could take on a half-integer, rather than integer value.[16] Heisenberg proposed a model in which the electrons move on two different ellipses on either side of the nucleus, in opposite directions. Under the influence of the mutual electron interaction, the angle φ between the major axes of the two orbits changes; Heisenberg introduced an additional quantum condition for this φ and proposed that the corresponding quantum number be half-integer:

$$\oint p_\varphi \mathrm{d}\varphi = n_\varphi h \qquad \text{where } n_\varphi = 1/2, 3/2, 5/2, \ldots \tag{4.6}$$

With this model and the ad hoc assumption of half-integer quantum numbers Heisenberg was able to obtain a value for the ground state of helium atom that was remarkably accurate: "I obtain an ionization potential of 24.6 Volt ... This agrees perfectly with the value of 24.5 Volt measured spectroscopically by Lyman" (Heisenberg to Sommerfeld, October 28th, 1922; quoted in Mehra and Rechenberg 1982, Vol. 2, p. 88).[17] Despite this success, both Pauli and Bohr were critical of the introduction of half-integer quantum numbers; whether it was because of their criticisms or his subsequent work with Max Born, Heisenberg never published his results.

Not only did the ground state of helium prove to be a stumbling block for the old quantum theory, but as Born and Heisenberg showed in 1923, the calculations of the excited states of helium proved even more devastating. The correspondence from this year gives a sense of the feeling of crisis that the failure of the helium atom precipitated. Heisenberg describes the results of his paper with Born in a letter to Pauli as follows:

[16] Although Heisenberg never published his half-integer approach to the helium atom, it is described in some detail in a letter to Sommerfeld on October 28, 1922. My discussion of Heisenberg's half-integer approach to the helium atom here follows that of Mehra and Rechenberg (1982) Vol. 2, Section II.3. Heisenberg had also tried to introduce half-integer quantum numbers to account for the "anomalous Zeeman effect" (the fact that the spectral lines of an atom in a magnetic field split in unexpected ways due to the not yet discovered electron spin); for a discussion of this see Massimi (2005), Chapter 2.

[17] With hindsight, Heisenberg's success here can be understood as largely coincidental. Semiclassical theorist Gregor Tanner and colleagues explain, "The good agreement between the results of Heisenberg's first perturbative calculations of the helium ground-state energy ... and the experimental values must be considered as accidental: Heisenberg's perturbative scheme was not appropriate ... and he did not know the role of Maslov indices in quantization conditions" (Tanner *et al.* 2000, p. 503). The correct semiclassical approach to the helium atom will be discussed in Section 5.3.

Born and I have now calculated the most general model of helium in all details ... and we find that the energy certainly comes out to be wrong ... The result appears to me – provided no error is hidden somewhere – to be very bad news for our present views; one will probably have to introduce totally new hypotheses – either new quantum conditions or assumptions to alter mechanics.

(Heisenberg to Pauli, February 19th, 1923; quoted in Mehra and Rechenberg 1982, Vol. 2, p. 93)

Born followed up on this letter with another letter to Bohr also emphasizing the negative implications that he saw this paper having for the old quantum theory:

A short time ago Heisenberg wrote to Pauli <then in Copenhagen> that, together with me, he had carried out a calculation of the terms of the excited helium atom ... Our result is quite negative. When one seeks out *all* the possible types of orbits of excited He by means of perturbation theory, then one finds no energy level (except the p-term) that agrees with the observed terms. Hence, even here, where one of the electrons is far removed, there must occur deviations from the laws of mechanics, or from the quantum rules. Now, I shall not pester you by discussing all the ways in which one might avoid this catastrophe.

(Born to Bohr, March 4th, 1923; quoted in BCW 2, p. 38)

As Born's letter reveals, the failure to obtain reasonable values for the excited states of helium – despite repeated and extensive theoretical efforts – was perceived as a catastrophe. There was a growing sense in the scientific community that the fault lay, not with a particular model or calculational scheme, but rather with the mechanics of the old quantum theory itself.

Three years later, after the advent of the new quantum theory, Born summarized the state of the old quantum theory with regard to the helium atom as follows:

[A]ttempts have been made to find the stationary orbits for the helium atom, and calculate its energy levels. The line of attack has been along two directions: Some investigators have considered the normal state of the helium atom (Bohr, Kramers, van Vleck), others the excited state, in which one electron is in the nearest orbit to the nucleus and the other revolves in a very distant orbit (van Vleck, Born and Heisenberg). Both calculations give incorrect results: The calculated energy of the normal state does not agree with experimental results (ionization energy of the normal helium atom), and the calculated term system for the excited states is different from that observed, qualitatively as well as quantitatively.

(Born 1926, pp. 47–8)

Although there were other problems plaguing the old quantum theory, such as the anomalous Zeeman effect, the failure of the helium atom had a certain symbolic significance. The historian of science Olivier Darrigol has cogently argued that, in Bohr's eyes, it was the real source of the crisis in 1923 (Darrigol 1992, p. 210).

On the heels of the failure of the helium atom followed the view of Born, Pauli, and Heisenberg that the concept of electrons following classical orbits was no longer appropriate or useful in the quantum realm. In 1924 Pauli wrote to Bohr:

> To me the most important question is: To what extent one can speak of well-defined trajectories of the electrons in the stationary states … In my opinion Heisenberg hits the truth precisely when he doubts that it is possible to speak of determinate trajectories.
>
> *(Pauli to Bohr, February 21st, 1924)*

A year later, Heisenberg introduced his matrix mechanics paper, and the idea of electron orbits would be banished from the new quantum theory.

It was not until 1929 that the Norwegian physicist Egil Hylleraas was able to show that the new quantum theory yielded a satisfactory value for the ground state of the helium atom. Hylleraas was set this problem by Born, who felt that the ground state of the helium atom remained an outstanding crucial test of the correctness of the new quantum theory (Hylleraas 1963, p. 425). In recalling the outcome of his research on the helium atom, Hylleraas notes,

> The end result of my calculations was a ground-state energy of the helium atom corresponding to an ionization energy of 24.35 eV which was greatly admired and thought of as almost a proof of the validity of wave mechanics, also, in the strict numerical sense. The truth about it, however, was, in fact, that its deviation from the experimental value by an amount of one-tenth of an electron volt was on the spectroscopic scale quite a substantial quantity and might well have been taken to be a disproof.
>
> *(Hylleraas 1963, p. 427)*

Hylleraas's subsequent calculations of the ground state were able to further close this gap, and attempts to improve the agreement between the quantum theoretical and experimental values of the ground state of helium lasted well into the 1950s.

4.2 The correspondence principle

Quantum theory is typically viewed as having made a revolutionary break from classical mechanics. Remarkably, however, Bohr saw the development from classical electrodynamics to the old quantum theory and then to the new quantum theory as forming one continuous development.[18] The linchpin of this continuity, in Bohr's view, is the correspondence principle. Upon the publication of Heisenberg's matrix mechanics paper, ushering in the new quantum theory, Bohr even went so far as to say that "the whole apparatus of the [new] quantum mechanics can be regarded as a precise formulation of the tendencies embodied in the correspondence principle"

[18] I do not mean to suggest that Bohr failed to recognize the important differences between these theories; rather, there is a continuity in the sense that he viewed the quantum theory as a *rational generalization* of the classical theory, as I shall argue in detail in Section 4.3.

(Bohr 1925, p. 852; BCW 5, p. 280). This claim is puzzling in light of the widespread belief that Bohr's correspondence principle is just the requirement that, in the limit of large quantum numbers, there should be an agreement between the predictions of quantum and classical mechanics. We encountered this view of the correspondence principle in Section 1.4 as a special case of Nickles' reductionism$_2$ relation.

Due to the efforts of historians and philosophers of science such as Nielsen (1976), Batterman (1991), Darrigol (1992), and Tanona (2002), there is now a growing recognition that Bohr's correspondence principle was *not* intended as some sort of requirement that classical mechanics be recovered out of quantum mechanics in the limit of high quantum numbers (despite many statements to the contrary in the physics literature). Reexamining precisely how the correspondence principle should instead be understood is the subject of this section.

The first germ of the correspondence principle, as Bohr himself reports (Bohr 1922), can be found in his 1913 lecture "On the constitution of molecules and atoms," although the term does not appear in his writings until 1920.[19] In the years before Bohr adopted the expression "correspondence principle" he used a locution quite similar to Paul Dirac's, namely that of seeing how closely one can trace the analogy between classical and quantum mechanics. For example, in 1918 Bohr writes, "It seems possible to throw some light on the outstanding difficulties by trying to trace the analogy between the quantum theory and the ordinary theory of radiation as closely as possible" (Bohr 1918, p. 4; BCW 3, p. 70).[20] In his later writing, however, Bohr explicitly rejects this view that the correspondence principle can be thought of as an analogy between the two theories. He writes,

> In Q.o.L. [Bohr 1918] this designation has not yet been used, but the substance of the principle is referred to there as a formal analogy between the quantum theory and the classical theory. Such expressions might cause misunderstanding, since in fact – as we shall see later on – this Correspondence Principle must be regarded purely as a law of the quantum theory, which can in no way diminish the contrast between the postulates and electrodynamic theory.
>
> *(Bohr 1924, fn. p. 22)*

The fact that Bohr refers to the correspondence principle as a *law* of quantum theory suggests, first, that he takes it to be a *universal* principle (not just applicable in a limited domain), and, second, that it is an essential part of quantum theory itself, not

[19] For a comprehensive history of the correspondence principle see Rud Nielsen's introduction to Volume 3 of *Niels Bohr Collected Works* (Nielsen 1976).

[20] Bohr later refers to this paper as "Q. o. L.," an abbreviation for the title "On the Quantum Theory of Line Spectra." Despite the prima facie similarity between Dirac's and Bohr's use of analogy, as we shall soon see, they in fact end up with quite distinct views of the relation between classical and quantum mechanics.

some sort of general methodological constraint coming from outside of quantum theory.

Recall that according to the old quantum theory, it is assumed that while the motion of an electron *within* a particular stationary state can still be described on the basis of classical mechanics, the radiation given off in a transition *between* stationary states (the "quantum jumps") cannot. The fundamental insight of Bohr's correspondence principle is that even these quantum transitions are determined in a surprising way by the classical description of the electron's motion.[21] In order to more clearly understand the substance of Bohr's correspondence principle as a rapprochement of the quantum and classical theories, it is helpful to first review more precisely how these theories differ.

Following Bohr, it is easiest to present the correspondence principle in the context of a simplified model of the atom as a one-dimensional system, where the electron is undergoing simply periodic motion. Classically, the trajectory of the electron is given by $x(t)$, which is the solution to Newton's equation of motion, and is periodic, which means it simply retraces its steps over and over again with a frequency ω, known as the fundamental frequency. Because the motion is periodic, the position of the electron can be represented by a Fourier series as follows:[22]

$$x(t) = C_1 \cos \omega t + C_2 \cos 2\omega t + C_3 \cos 3\omega t + \cdots \tag{4.7}$$

Each of these terms in the sum is known as a harmonic, and the τth harmonic is given in terms of the time t, an amplitude C_τ, and a frequency ω_τ, which is an integer multiple of the fundamental frequency $\omega_\tau = \tau\omega$ (these multiples of the fundamental frequency are referred to as the "overtones"). According to classical electrodynamics, the frequencies of the radiation emitted by this atom should just be given by the frequencies in the harmonics of the motion: ω, 2ω, 3ω, etc.; hence the spectrum of this classical atom should be a series of discrete evenly spaced lines.[23]

According to Bohr's old quantum theory, by contrast, the radiation is not a result of the accelerated motion of the electron in its orbit, but rather of the electron jumping from one stationary state to another; and rather than giving off all of the harmonic "overtones" together, only a single frequency ν, is emitted, and the value

[21] As shall become clearer below, it is the distance of the jump (the change in quantum number) – not the time at which the jump occurs – that is determined by the classical motion. I shall argue that the more familiar asymptotic agreement of quantum and classical frequencies for large quantum numbers is in fact a consequence of Bohr's correspondence principle, but not the correspondence principle itself.

[22] A Fourier series, recall, is a way of representing any periodic function $F(x)$ in terms of a weighted sum of sinusoidal components (e.g., sines and cosines).

[23] In presenting the physics behind Bohr's correspondence principle, I have benefited from an excellent article by Fedak and Prentis (2002). As Fedak and Prentis (2002) explain, because the electron is radiating energy away, its motion will not be strictly periodic, but rather a spiral to the nucleus; if, however, the initial energy of the electron is large compared to energy being lost, then this loss can be neglected and the motion is well-approximated as being periodic, and hence its spectrum will be discrete (p. 333).

of that frequency is given by the Bohr–Einstein frequency condition (Equation 4.1). The spectral lines are built up by a whole ensemble of atoms undergoing transitions between different stationary states, and these spectral lines, though they exhibit a pattern of regularity, are not evenly spaced – except in the limit of large quantum numbers.

It was by investigating this limit of large quantum numbers that Bohr uncovered the remarkable fact that, despite these striking differences between the quantum and classical theories, there is nonetheless a deep relation between the quantum frequency ν, and the harmonic components of the classical motion. More specifically, Bohr considers the radiation that is emitted in the transition between two stationary states, labeled by quantum numbers n' and n'', in the case where these quantum numbers are large compared to the difference between their values. Looking ahead, we can label the difference between the n'th stationary state and the n''th stationary state by τ (e.g., if the electron jumps to the nearest stationary state, $\tau = 1$; if it jumps two stationary states away, $\tau = 2$; and so on).[24] In this high n limit, Bohr discovered that the frequency of radiation $\nu_{n' \to n''}$ emitted in a quantum jump of difference τ from state n' to n'' is equal to the frequency in the τth harmonic of the classical motion in the n' stationary state, namely

$$\nu_{n' \to n''} = \tau\omega, \quad \text{where } n' - n'' = \tau. \tag{4.8}$$

It is this equality between the quantum frequency and one component of the classical frequency, in the limit of large quantum numbers, that was the "clue" that led Bohr to what he calls the "law" of the correspondence principle.

In Bohr's own words, the correspondence principle can be characterized as follows:

> [T]he possibility of the occurrence of a transition, accompanied by radiation, between two states of a multiply periodic system, of quantum numbers for example $n_1', \ldots, n_u'$ and $n_1'', \ldots, n_u''$, is conditioned by the presence of certain harmonic components in the expression given by … [the Fourier series expansion of the classical electron motion]. The frequencies $\tau_1\omega_1 + \ldots + \tau_u\omega_u$ of these harmonic components are given by the following equation
>
> $$\tau_1 = n_1' - n_1'', \ldots, \tau_u = n_u' - n_u''. \tag{4.9}$$
>
> We, therefore, call these the "corresponding" harmonic components in the motion, and the substance of the above statement we designate as the "Correspondence Principle."
>
> *(Bohr 1924, p. 22; BCW 3, p. 479)*

I want to argue that, as we see in this passage, Bohr's correspondence principle is not the asymptotic agreement of quantum and classical frequencies, but rather what we

[24] Although classically τ specifies a particular harmonic component of the classical motion and quantum mechanically τ specifies the change in the quantum number in a particular jump, the fact that these physically different 'τ' are numerically equal is the deep insight of the correspondence principle (Fedak and Prentis 2002, p. 335).

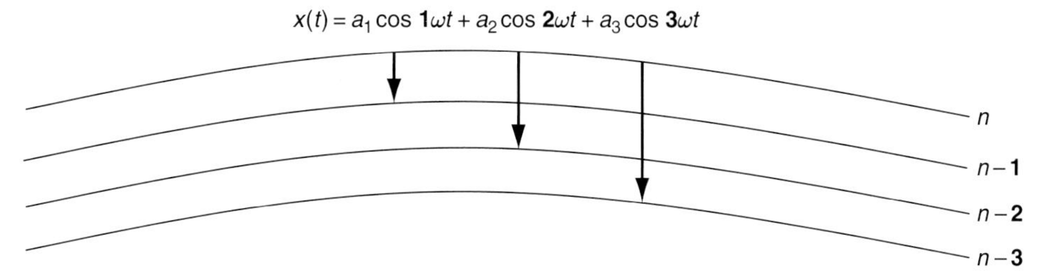

Figure 4.2 The correspondence principle is Bohr's insight that each allowed transition between stationary states (labeled by n) corresponds to one harmonic component of the classical motion. [Redrawn, with permission, from Figure 3 of Fedak and Prentis (2002). Copyright 2002, American Association of Physics Teachers.]

might call "*Bohr's selection rule*." This selection rule states that the transition from a stationary state n' to another stationary state n'' is allowed *if and only if* there exists a τth harmonic in the classical motion of the electron in the stationary state; if there is no τth harmonic in the classical motion, then transitions between stationary states whose separation is τ are not allowed quantum mechanically.[25] The essence of Bohr's correspondence principle is depicted in Figure 4.2.

It is worth taking a brief detour from textual exegesis here to illustrate Bohr's selection rule by considering the following simplified example.[26] Suppose that the solution to Newton's equation, $F = m\ddot{x}$, and the quantum condition $\oint p dx = nh$ is[27]

$$x(t, n) = n \cos \sqrt{n}t + \sqrt{n} \cos 3\sqrt{n}t, \tag{4.10}$$

which is the Fourier decomposition of the classical periodic motion of the electron in an allowed stationary state n. For this stationary state n, the fundamental frequency (i.e., periodicity of the electron motion) is $\omega = \sqrt{n}$. Note that there are only the

[25] Although the authors that I mentioned at the beginning of this section, who have gone a long way towards a correct characterization of the correspondence principle (CP), come close to the view I am defending here, none of them quite define the CP in this way. Tanona, for example, takes the CP to be a connection between the atomic *spectrum* (radiation) and the classical orbital mechanics (rather than the *transitions* and the orbital mechanics) (Tanona 2002, p. 60). Darrigol takes the CP to be a connection between the *amplitude* of the harmonic components and the *probabilities* of the transitions (Darrigol 1992, p. 126). I want to argue instead that what these authors are identifying are *correspondences* which are consequences of, or applications of, the CP, but not the CP itself. Batterman, who in particular has rightly emphasized that the CP is an *explanation* or *justification* for the asymptotic agreement (not an asymptotic agreement itself) seems to be somewhere in between Tanona and myself in identifying the CP as a correspondence between "radiative processes" and periodic motions of the electron (Batterman 1991, p. 203).

[26] From Fedak and Prentis (2002, p. 337).

[27] One substitutes the Fourier series representation of the solution $x(t)$ to Newton's equation into the quantum condition to obtain a "quantized" Fourier series representation of the solution $x(t,n)$ of the following form:

$$x(t, n) = \sum_{k=1}^{\infty} C_k(n) \cos k\omega(n)t.$$

first ($\tau = 1$) and third ($\tau = 3$) harmonics present in the classical motion. According to Bohr's selection rule, this means that there can only be quantum jumps between stationary states that are one or three stationary states apart. So, for example, there can be transitions from the $n = 100$ stationary state to the $n = 99$ or $n = 97$ stationary states; but there cannot be transitions from the $n = 100$ stationary state to the $n = 98$ stationary state, because there is no second harmonic in the classical electron orbit.

My claim, then, is that the correspondence principle is the following statement: which quantum transitions are allowable is *determined by* which harmonics are present in the classical motion. As Bohr notes, the correspondence principle provides

> an immediate interpretation of the apparent capriciousness, involved in the application of the principle of combination of spectral lines, which consists in the circumstance, that only a small part of the spectral lines, which might be anticipated from an unrestricted application of this [Rydberg–Ritz combination] principle, are actually observed in the experiments.
> *(Bohr 1921b unpublished; BCW 4, p. 150)*

In addition to explaining the capriciousness of the spectral lines, the correspondence principle also leads to an explanation of several "correspondences," or what Bohr sometimes calls "applications of the correspondence principle" (in German, "*Anwendung des Korrespondenzprinzips*") (Bohr 1921a, xii; BCW 3, p. 331).[28] One of these applications of the correspondence principle is the asymptotic agreement between the quantum radiation frequency and the classical frequency of the corresponding harmonic component, which was discussed above. Another application of the correspondence principle that Bohr often discusses is the correspondence between the intensities of the spectral lines, which are given by transition probabilities, and the *amplitude* of the corresponding harmonic in the classical motion.

Classically the intensity of a frequency $\tau\omega$ depends on the amplitude C_τ. Quantum mechanically, however, the intensity depends on how probable a particular transition is between two stationary states; that is, more photons will be given off at a frequency given by highly probable transitions between stationary states and hence that spectral line will be brighter, whereas if a transition between two stationary states is unlikely to occur, then fewer photons will be given off at that frequency and that spectral line will be fainter. As Bohr notes, the quantum-mechanical account depends on using the notion of probabilities for transitions, which Einstein introduced in his theory of heat radiation, and this "raises the serious question of whether we must rest content with statements of probabilities for individual processes. As matters stand at present, we are so far from being able to give a real description of these processes that we may well assume that Einstein's mode of treatment may actually be the most appropriate" (Bohr

[28] It is perhaps helpful to note that when Bohr is in his "context of discovery" mode, he talks about these asymptotic correspondences as "clues" to the CP, whereas when he is in his "context of justification" mode he refers to them as "applications."

1922 unpublished lecture; BCW 4, p. 348).[29] Despite these striking physical differences between the classical and quantum intensities, Bohr notes that there is, nonetheless, another direct asymptotic correspondence:

> [A] relation, as that just proved for the frequencies, will, in the limit of large n, hold also for the intensities of the different lines in the spectrum. Since now on ordinary electrodynamics the intensities of the radiations corresponding to different values of τ are directly determined from the coefficients C_τ in $[x(t) = \sum C_\tau \cos 2\pi\,(\tau\omega t + c_\tau)]$ we must therefore expect that for large values of n these coefficients will on the quantum theory determine the probability of spontaneous transition from a given stationary state for which $n = n'$ to a neighboring state for which $n = n'' = n' - \tau$.
>
> *(Bohr 1918, p. 15; BCW 3, p. 81)*

In other words, in the limit of large n the probability of a transition between two stationary states separated by τ is given by the (square of the) amplitude of the τth harmonic component of the classical motion.[30] Thus in the limit of large n the amplitudes of the harmonic components of the electron's classical orbit can be used to calculate the intensities of the spectral lines.[31] Bohr notes that these correspondences can also be used to determine the polarization of the photon emitted in the transition.[32]

It is important to recognize, however, that none of these particular correspondences, which, in the limit of large n, allow for a direct calculation of various quantum quantities from the classical harmonic components, are themselves the correspondence principle. Rather they can be used alternatively as inductive evidence for the correspondence principle, or, once the correspondence principle is established, as applications or consequences of the correspondence principle.[33] The

[29] Given the usual understanding of the Einstein–Bohr debate (see Section 3.5) this is an interesting twist in that it is *Bohr* who is uncomfortable with *Einstein's* introduction of probabilities into quantum mechanics!

[30] More precisely the quantum transition probability $P_{n' \to n''}$ is given by

$$P_{n' \to n''} = \frac{e^2 \tau^2 \omega^3(n)}{12\pi\varepsilon_0 \hbar c} (C_\tau(n))^2$$

where e is the charge of the electron, ε_0 the permittivity of free space, and c the speed of light. For further details see Fedak and Prentis (2002, p. 335).

[31] Note that for both the frequencies and intensities, one can only speak of a *statistical* asymptotic agreement, since in a quantum transition only one photon is emitted with one frequency.

[32] For example, Bohr notes "Further as regards the state of polarisation of the radiation corresponding to the various transitions we shall in general expect an elliptical polarisation in accordance with the fact, that in the general case the constituent harmonic vibrations of a multiple-periodic motion possess an elliptical character ..." (Bohr 1921b, unpublished lecture; BCW 4, p. 150).

[33] Although Darrigol's interpretation differs slightly from the one I am arguing for here, he also notes these inductive and deductive uses of the correspondence principle when he writes, "The precise expression and scope of the CP depended on the assumptions made about the electronic motion. Whenever this motion was a priori determined, the 'correspondence' aided in *deducing* properties of emitted radiation. In the opposite case, characteristics of the electronic motion could be *induced* from the observed atomic spectra. This ambiguity made the CP a very flexible tool that was able to draw the most from the permanent inflow of empirical data" (Darrigol 1992, p. 83).

correspondence principle is a more general relation underlying these various particular correspondences. Once again, Bohr's correspondence principle states that only those quantum transitions are allowed that have a corresponding harmonic component in the classical electron orbit. This interpretation of the correspondence principle as Bohr's selection rule, rather than any sort of statistical asymptotic agreement involving frequencies or intensities, allows us to make sense of three prima facie puzzling claims by Bohr: first, that the correspondence principle applies to *small n* as well as large n (it is not just an asymptotic relation); second, that the correspondence principle is a *law of quantum theory*; and third, that the essence of the correspondence principle survives in the new matrix mechanics.

For those who interpret the correspondence principle as the asymptotic agreement between quantum and classical frequencies for large n, there has always been the troubling fact that Bohr claims the correspondence principle also applies to small n. We see this, for example, in Bohr's discussion of the well-known red and green spectral lines of the Balmer series in the visible part of the hydrogen spectrum. The red spectral line (which really is red at a wavelength of around 656 nm) is typically labeled H_α, and is the result of radiation emitted in the jump from the $n = 3$ to $n = 2$ stationary state. The green line (labeled H_β with a wavelength of around 486 nm) is a result of the electron in a hydrogen atom jumping from the $n = 4$ to $n = 2$ stationary state. Regarding these low-quantum-number transitions Bohr writes,

> We may regard H_β as the octave of H_α, since H_β corresponds to a jump of 2 and H_α to a quantum jump of 1. It is true that H_β does not have twice the frequency of H_α, but it corresponds to the octave. This relationship we call the "correspondence principle." To each transition there corresponds a harmonic component of the mechanical motion.
>
> *(Bohr 1922 unpublished lecture; BCW4, p. 348)*

In other words, although the "frequency correspondence" does not hold for these low quantum numbers (nor can the intensities of these lines be calculated directly from the classical amplitudes via the "intensity correspondence"), the more general *correspondence principle* does hold; specifically, these $\tau = 1$ and $\tau = 2$ transitions are allowed because there is, in the Fourier decomposition of the electron's classical orbit, a first and second harmonic component. Or again, when Bohr is generalizing from the asymptotic intensity correspondence, he writes,

> This peculiar relation suggests a *general law for the occurrence of transitions between stationary states*. Thus we shall assume that even when the quantum numbers are small the possibility of transition between two stationary states is connected with the presence of a certain harmonic component in the motion of the system.
>
> *(Bohr 1920, p. 28; BCW 3, p. 250)*

It is important to note that Bohr's selection rule applies to *all* quantum transitions, not just those transitions in the limit of large quantum numbers. Hence, interpreting the correspondence principle as Bohr's selection rule allows us to straightforwardly make sense of these claims that the correspondence principle applies to small quantum numbers as well.[34]

The above quotation also helps us make sense of Bohr's claim that the correspondence principle is a law of quantum theory. It is a *law* because it is a universal (i.e., applying to all n) restriction on the allowed quantum transitions. To understand why it is a law *of quantum theory* (as opposed to classical electrodynamics) it is helpful to consider Bohr's following remarks:

> [T]he occurrence of radiative transitions is conditioned by the presence of the corresponding vibrations in the motion of the atom. As to our right to regard the asymptotic relation obtained as the intimation of a general law of the quantum theory for the occurrence of radiation, as it is assumed to be in the Correspondence Principle mentioned above, let it be once more recalled that in the limiting region of large quantum numbers there is no wise a question of a gradual diminution of the difference between the description by the quantum theory of the phenomena of radiation and the ideas of classical electrodynamics, but only an asymptotic agreement of the statistical results.
>
> *(Bohr 1924, p. 23; BCW 3, p. 480)*

In this passage we see that Bohr takes quantum mechanics to be a universal theory. Despite the statistical agreement of results in this limit, the physics behind the meanings of "frequency" and "intensity," for example, remains different, and Bohr is insistent that it is the quantum account that is the strictly correct one – even in this high n or "classical" limit. Hence when Bohr discovered that the allowable quantum transitions are those for which there is a corresponding harmonic in the classical motion, what he had discovered was something about quantum theory.

So far we have seen how interpreting the correspondence principle as Bohr's selection rule, rather than as some sort of asymptotic agreement between classical and quantum frequencies, can help us make sense of Bohr's claims that the correspondence principle applies just as well to small quantum numbers as it does to large, and that the correspondence principle is a law of quantum theory. The third puzzle concerning the correspondence principle is Bohr's claim that the new quantum theory, in its entirety, can in some sense be thought of as a formalization of the correspondence principle. In describing Heisenberg's new matrix mechanics Bohr writes,

[34] Bohr does think, however, that, from this fact that the CP applies universally (to all n), it should also therefore follow that there is a more general and complicated correspondence relation holding between both quantum and classical frequencies and between quantum transition probabilities and classical amplitudes, which "hide themselves" in the low-quantum-number regime. He describes the extension or generalization of these frequency and intensity correspondences to all n in Bohr (1924, p. 24; BCW 3, p. 481); for a modern treatment see Fedak and Prentis (2002, p. 336).

It operates with manifolds of quantities, which replace the harmonic oscillating components of the [classical] motion and symbolise the possibilities of transitions between stationary states in conformity with the correspondence principle. These quantities satisfy certain relations which take the place of the mechanical equations of motion and the quantisation rules … The classification of stationary states is based solely on a consideration of the transition possibilities, which enable the manifold of these states to be built up step by step. In brief, the whole apparatus of the quantum mechanics can be regarded as a precise formulation of the tendencies embodied in the correspondence principle.

(Bohr 1925, p. 852; BCW 5, p. 280)

Bohr's claim here is that Heisenberg's matrix elements ("manifolds of quantities") are the counterpart to the harmonic components of the classical motion, and the way that those matrix elements symbolize the transition probabilities is in accordance with the correspondence principle. Assessing this claim of Bohr's, however, requires a brief detour into Heisenberg's matrix mechanics paper.

When one takes a closer look at Heisenberg's 1925 paper it is perhaps surprising that his well-known declared strategy of building up the new quantum theory on the basis of observable quantities alone turns out to be in no way incompatible with his lesser-known declared strategy of trying to "construct a quantum-mechanical formalism corresponding as closely as possible to that of classical mechanics" (Heisenberg [1925] 1967, p. 267).[35] This is in striking contrast to Heisenberg's later recollections, which we examined in Chapter 2, in which he declares that the development of quantum mechanics required him to "cut the branch" on which he was sitting, and make a "clean break" with classical mechanics. Throughout the 1925 paper, Heisenberg begins by writing down how a problem would be set up and solved classically, and then notes the minimum possible changes that are required to set up and solve these equations in the quantum case. In this perhaps minimal sense Heisenberg is indeed following the spirit underlying the correspondence principle, which Bohr describes as follows: "The correspondence principle expresses the tendency to utilise in the systematic development of the quantum theory every feature of the classical theories in a rational transcription appropriate to the fundamental contrast between the [quantum] postulates and the classical theories" (Bohr 1925, p. 849; BCW 5, p. 277).[36] However, one can see the correspondence principle at work in Heisenberg's matrix mechanics paper even more directly.

Characterizing the radiation in terms of observables only, for Heisenberg, means eliminating all reference to the position and period of revolution of the electron, and instead finding the quantum-mechanical expressions for the frequencies (which, as in the old quantum theory, are given by the Einstein–Bohr frequency condition) and

[35] Darrigol has cogently argued that "contrary to common belief, the reduction to observables did not directly contribute to Heisenberg's discovery [of quantum mechanics]" (Darrigol 1997, fn. p. 558; see also Darrigol 1992, pp. 273–5).

[36] Earlier I referred to this broad methodological formulation of the correspondence principle as the 'general correspondence principle'.

the transition amplitudes. Heisenberg notes that quantum mechanically the amplitudes will be complex vectors, and while classically the amplitude is given by[37]

$$\mathrm{Re}\left[C_\tau(n)\mathrm{e}^{\mathrm{i}\omega(n)\tau t}\right], \tag{4.11}$$

quantum mechanically the amplitude is given by

$$\mathrm{Re}\left[C(n, n-\tau)\mathrm{e}^{\mathrm{i}\omega(n,\, n-\tau)t}\right]. \tag{4.12}$$

Although classically these phases give the frequencies of the radiation, Heisenberg notes that

> [a]t first sight the phase contained in C would seem to be devoid of physical significance in quantum theory, since in this theory frequencies are in general not commensurable with their harmonics. However, we shall see presently that also in quantum theory the phase has a definite significance which is analogous to its significance in classical theory.
>
> *(Heisenberg [1925] 1967, p. 264)*

As Bohr's frequency correspondence shows, the quantum frequencies are only commensurable with their classical harmonics in the limit of large quantum numbers. Although quantum mechanically the ω in the phase are not typically equal to the frequencies ν, Heisenberg notes that they do have a physical significance that is *analogous*. To see what this physical significance is, we need to examine how Heisenberg represents the quantum analog of the Fourier series decomposition of the classical electron trajectory in a stationary state n.

Recall that for classical periodic motion one can represent the electron trajectory as a Fourier series:

$$x(n, t) = \sum_{\tau=-\infty}^{+\infty} C_\tau(n)\mathrm{e}^{\mathrm{i}\omega(n)\tau t}. \tag{4.13}$$

Regarding this classical decomposition, Heisenberg writes,

> A similar combination of the corresponding quantum-theoretical quantities seems to be impossible in a unique manner and therefore is not meaningful, in view of the equal weight of the variables n and $n - \tau$. However, one may readily regard the ensemble of quantities
>
> $$C(n, n-\tau)\mathrm{e}^{\mathrm{i}\omega(n,\, n-\tau)t}$$
>
> as a representation of the quantity $x(t)$.
>
> *(Heisenberg [1925] 1967, p. 264)*

[37] I have tried to keep as close to Heisenberg's original notation as possible, changing only his 𝔄 for C and α for τ to be consistent with my earlier notation.

In other words, what Heisenberg is essentially doing is taking the harmonics of the classical motion of the electron in its orbit and turning them into complex elements of a matrix, with the n and $n - \tau$ as the indices labeling those matrix elements. This is the way that the new quantum theory incorporates Bohr's correspondence principle law that only those quantum transitions are allowed that have a corresponding classical harmonic. If there is no τth harmonic in the classical motion, then the matrix element labeled n, $n - \tau$ will be zero, meaning that particular transition probability is zero. Heisenberg then goes on to work out the multiplication rules for these matrices and notes that they are noncommutative.

After figuring out the quantum dynamics in the second section of his paper, Heisenberg then goes on in the third section to apply this new matrix mechanics to the simple example of an anharmonic oscillator, $\ddot{x} + \omega_0^2 x + \lambda x^3 = 0$. Here we see again quite explicitly what is essentially an application of Bohr's correspondence principle. Heisenberg begins by writing down the Fourier decomposition for the classical trajectory $x(t)$ and notes, "Classically, one can in this case set $x = a_1 \cos \omega t + \lambda a_3 \cos 3\omega t + \lambda^2 a_5 \cos 5\omega t + \cdots$; quantum-theoretically we attempt to set by analogy $a(n, n-1) \cos \omega (n, n-1)t$; $\lambda a(n, n-3) \cos \omega (n, n-3)t$; ..." (Heisenberg [1925] 1967, p. 272). Note that for this example, there are only the odd harmonics in the classical motion: 1ω, 3ω, 5ω, etc. Hence, by Bohr's correspondence principle, the only allowed quantum transitions are those that jump $\tau = 1, 3, 5$, etc. stationary states. This gets incorporated into Heisenberg's matrix mechanics in that the matrix elements, which give the quantum transition amplitudes, $a(n, n - \tau)$, are precisely the quantum analogues of these odd harmonics in the classical motion. Because there are no even harmonics in the motion of the *classical* anharmonic oscillator, the corresponding matrix elements $x_{n,\, n-2}$, $x_{n,\, n-4}$, $x_{n,\, n-6}$, etc. of the *quantum* anharmonic oscillator will be zero, meaning the transition probabilities between these states are zero. Hence, as this detour into Heisenberg's 1925 matrix mechanics paper shows, if we interpret Bohr's correspondence principle as Bohr's selection rule, then there is a straightforward sense in which the correspondence principle *does* survive in the new quantum theory as Bohr claimed.[38]

Although there is a sense in which Heisenberg's matrix mechanics incorporates Bohr's correspondence principle into the new formalization, Heisenberg himself was never entirely comfortable with the correspondence principle, and certainly never embraced it as a law of the quantum theory. He makes this clear in his book *The Physical Principles of the Quantum Theory*:

[38] As Darrigol (1997) rightly emphasizes, although the CP is incorporated into matrix mechanics, there is no longer a literal interpretation of the electron motion in the stationary state; instead it is a "symbolic use, in which the space-time relations are completely lost" (p. 559).

> It is true that an ingenious combination of arguments based on the correspondence principle can make the quantum theory of matter together with a classical theory of radiation furnish quantitative values for the transition probabilities ... Such a formulation of the radiation problem is far from satisfactory, however, and easily leads to false conclusions.
>
> *(Heisenberg 1930, p. 82)*

Before showing how correspondence arguments might lead to false conclusions, however, Heisenberg shows (à la Bohr) how to interpret the elements of his quantum matrix via the correspondence principle as being related to the Fourier coefficients of the classical motion.[39] He concludes, however, that "[i]t must be emphasized that this [correspondence] is a purely formal result; it does not follow from any of the physical principles of quantum theory" (Heisenberg 1930, p. 83). In other words, the correspondence is to be interpreted as a purely mathematical result, not as any deep connection between the quantum and classical theories, and certainly not as a principle of the quantum theory itself.[40] As we saw in Chapter 2, Heisenberg wants to view quantum mechanics as a closed theory, which is an axiomatic system complete in itself, not one that in any way depends on classical mechanics.

Heisenberg's views on the correspondence principle were likely influenced by Sommerfeld, who had been his dissertation advisor. As Darrigol has recounted in detail, Sommerfeld was never comfortable with Bohr's correspondence principle, and only begrudgingly admitted its fertility (Darrigol 1992, pp. 138–45). Sommerfeld preferred to view quantum theory as a self-contained set of formal rules, and wanted to derive the selection rules "by a remarkably rigorous manner of deduction, reminiscent of the incontrovertible logic of numerical calculations" (Sommerfeld [1919] 1923, pp. 265–6). By contrast, he disparagingly referred to Bohr's correspondence principle approach to the selection rules as follows: "Bohr has discovered in his *principle of correspondence* a magic wand (which he himself calls a formal principle), which allows us immediately to make use of the results of the classical wave theory in the quantum theory" (Sommerfeld [1919] 1923, p. 275).[41]

[39] The example that Heisenberg gives does not, in fact, show a problem with the correspondence principle itself; rather, it shows a problem with more general heuristic arguments based on the classical theory. Moreover, when those heuristic arguments are altered in such a way as to bring them into accordance with the CP, then the correct quantum-mechanical account once again obtains – a vindication of the CP!

[40] I think this debate between Bohr and Heisenberg about the significance of the correspondence principle in an interesting way parallels the debate, which will briefly be reviewed at the beginning of Chapter 6, between Batterman (2002; 2005) on one hand, and Belot (2005) and Redhead (2004) on the other, about the significance of appeals to classical structures in explaining quantum phenomena. Very roughly, Heisenberg, Redhead, and Belot want to view quantum mechanics as complete in itself, and any purported appeals to classical structures are really just appeals to abstract mathematics ("formal results"). Bohr and Batterman, by contrast, interpret such correspondences as indicating a deep relation between classical and quantum mechanics – showing that quantum mechanics in some sense depends on classical mechanics.

[41] Both of these Sommerfeld quotations are given in Darrigol (1992, p. 140).

Although Sommerfeld would later admit that the correspondence principle reveals an important relation between the quantum and classical theories, he describes this relation as one that continued to be "a source of distress" for him (Sommerfeld to Bohr November 11, 1920; BCW 3, p. 690).

For Bohr, the correspondence principle provides a deep link between classical mechanics, the old quantum theory, and the new quantum mechanics, tying all three of these theories together. It is interesting to note that the first occurrence of the expression "correspondence principle" is also the first occurrence of Bohr's claim that quantum theory is a "rational generalization" of the classical theory. As Bohr explains, despite the fundamental break between the quantum and classical theories of radiation,

> there is found, nevertheless, to exist a far-reaching *correspondence* between the various types of possible transitions between the stationary states on the one hand and the various harmonic components of the motion on the other hand. This correspondence is of such a nature, that the present theory of spectra is in a certain sense to be regarded as a rational generalization of the ordinary theory of radiation.
>
> *(Bohr 1920, p. 24; BCW 3, p. 246)*

As we shall see, Bohr maintained this view from 1920 onwards, taking it to apply not only to the old quantum theory, but to the new quantum mechanics as well. Understanding precisely what it might mean to call quantum mechanics a rational generalization of classical mechanics is the subject of the next section.

4.3 Quantum theory as a rational generalization

One of the least discussed and least understood aspects of Bohr's philosophy is his claim that quantum mechanics is a "rational generalization" of classical mechanics. In his own words, "quantum mechanics may be regarded in every respect as a generalization of the classical physical theories" (Bohr [1929] 1934, p. 4; BCW 6, p. 282). The centrality of the rational generalization thesis to Bohr's philosophy is evidenced by the fact that it is a point that appears repeatedly in his writings throughout his career, appearing in the context of both the old and new (post-1925) quantum theories. An understanding of this thesis is essential for an adequate account of Bohr's philosophy, and as I will show below, this thesis is closely intertwined with his better known views on the correspondence principle, complementarity, and the indispensability of classical concepts.

The foundation of Bohr's belief that the quantum theories are a rational generalization of the classical theories is the correspondence principle.[42] The correspondence principle, recall, is Bohr's discovery that, despite the fundamentally different physical

[42] The central importance of the correspondence principle for Bohr's rational generalization thesis was not yet adequately appreciated in Bokulich and Bokulich (2005).

mechanisms behind the classical and quantum accounts of radiation, there is nonetheless a deep connection between the harmonics in the classical motion and the allowed quantum transitions; moreover, this close connection is still preserved in Heisenberg's new matrix mechanics formulation of quantum theory. Early on Bohr emphasizes the importance of the correspondence principle in establishing the rational generalization thesis as follows:

> This fact [of the asymptotic connection between the spectrum and motion] leads one to regard the occurrence of every transition between stationary states ... as depending on the presence of a corresponding harmonic oscillation in the motion of the system. This viewpoint leads not only to the development of a general principle (the correspondence principle) ... but it makes it possible to regard the quantum theory as a rational generalization of the conceptions underlying the usual [classical] theory of radiation.
>
> *(Bohr 1921a; BCW 3, p. 331)*

The proper understanding of Bohr's correspondence principle as a law of quantum theory, rather than an asymptotic agreement of predictions, suggests that his rational generalization thesis should also be understood as a more substantive claim than simply the view that classical mechanics should be recovered from quantum mechanics in some limit.

Although Bohr sees Planck's discovery of the quantum of action as leading to the need for a fundamental revision in physics, in many ways he is more of a continuity theorist than a revolutionary. He is a continuity theorist in the sense that he tries to maintain and emphasize those features of the predecessor theory that are preserved in the transition to the successor theory. On Bohr's view, "The problem with which physicists were confronted was therefore to develop a rational generalization of classical physics, which would permit the harmonious incorporation of the quantum of action" (1958b, p. 309; BCW 7, p. 389). Note that Bohr's aim here is one of reconciliation – of bringing the classical and quantum theories together into a rational and consistent whole –, and the point of Bohr's rational generalization thesis is to explain precisely what the nature and extent of this continuity is. It is important to recognize that, for Bohr, this emphasis on continuity was not just a heuristic for theory construction, but was also an essential part of the proper understanding and interpretation of quantum theory.

The key to a harmonious incorporation of the quantum postulate into classical mechanics is determining the proper scope and applicability of classical concepts. One of the central lessons of the new quantum theory for Bohr is that not all classical concepts can be simultaneously applied to a given experimental situation. The answer to where and when certain classical concepts can be applied is to be found in his viewpoint of complementarity. He explains,

> [T]he indivisibility of the quantum of action ... forces us to adopt a new mode of description designated as *complementary* in the sense that any given application of classical concepts

precludes the simultaneous use of other classical concepts which in a different connection are equally necessary for the elucidation of the phenomena.

(Bohr [1929] 1934, p. 10; BCW 6, p. 288)

While classically one can simultaneously apply all relevant classical concepts to a given physical system, quantum mechanically one can apply only half of the relevant classical concepts; or, more precisely, one can simultaneously apply the concepts associated with complementary observables only up to the degree of accuracy permitted by the uncertainty principle. Which concepts apply is determined by the concrete experimental situation in which the physical system is being investigated.

Although Bohr saw complementarity as a straightforward consequence of Heisenberg's uncertainty principle, it was not an interpretation that either Heisenberg or Dirac shared.[43] In Heisenberg's 1963 interview with Kuhn, he explains how he and Bohr differed fundamentally on this point:

[F]or me it was clear that ultimately there was no dualism and after all, we had a closed mathematical scheme ... I was always a bit upset by this tendency of Bohr of putting it into a dualistic form saying, "here we have the waves and we can work around with the waves and here we have the particles and we shall play around with both."

(Heisenberg 1963, July 5th, p. 11)

As we saw in Chapter 2, Heisenberg's view of classical and quantum mechanics as distinct closed theories is incompatible with the sort of dualism of classical descriptions at the heart of Bohr's complementarity. Dirac similarly rejected Bohr's complementarity, though for different reasons than Heisenberg. When Kuhn asks Dirac what his views are on complementarity, Dirac responds, "I don't altogether like it. In the first place it is rather indefinite ... I feel that the last word hasn't been said yet on the relationship between waves and particles. When it has been said, people's ideas of complementarity will be different" (Dirac 1963, May 14th, p. 10). As we saw in the last chapter, Dirac was not convinced of the fundamental status of the uncertainty principle, and hence any philosophy of complementarity built on this relation is premature.

In some ways, Bohr's "generalization" of classical mechanics to the quantum context might better be described as a *restriction* of classical mechanics. The restriction in question, however, is not one of the domain of applicability of classical mechanics to the domain of large quantum numbers. Bohr is not simply referring to the uncontroversial point that classical mechanics, while no longer universal, nonetheless continues to provide an empirically adequate description of large-scale phenomena. Rather, Bohr's generalization thesis can be understood as a restriction

[43] Although it was Heisenberg's uncertainty principle that led Bohr to the viewpoint of complementarity, Bohr understood complementarity to be a much broader epistemological framework applicable to many other fields, such as biology.

in the following sense: "[I]t is the combination of features which are united in the classical mode of description but appear separated in the quantum theory that ultimately allows us to consider the latter as a natural generalization of the classical physical theories" (Bohr [1929] 1934, p. 19; BCW 6, p. 297). To put the point more bluntly, one might say that quantum mechanics just *is* a restriction of classical mechanics in accordance with the viewpoint of complementarity.

Nonetheless, Bohr takes quantum theory to be a generalization, not a restriction, of classical mechanics. He tries to explain the sense in which it is more general as follows: "In representing a generalization of classical mechanics suited to allow for the existence of the quantum of action, quantum mechanics offers a frame sufficiently wide to account for empirical regularities which cannot be comprised in the classical way of description" (Bohr 1948, p. 316; BCW 7, p. 334). In other words, quantum mechanics is a generalization in the sense that it is an extension of the classical theory that, in addition to the usual classical phenomena, allows for the incorporation of a fundamental unit of action, $\hbar$, and the new phenomena that Bohr sees this quantum of action bringing about.

It is important to emphasize that Bohr takes this generalization of classical mechanics to be *rational*. His insistence on the rationality of this enterprise is perhaps surprising, since most physicists and historians have viewed Bohr's blending of classical and quantum ideas as being – at best – "clever bricolage."[44] In his writings, however, one can see that the issue of consistency was never far from Bohr's mind. As we shall see next, Bohr understands a rational quantum mechanics to be one that maximally incorporates classical concepts, suitably reinterpreted, in a consistent manner.

4.4 The indispensability of classical concepts

Bohr's thesis that quantum mechanics is a rational generalization of classical mechanics is closely intertwined with his infamous doctrine of the indispensability of classical concepts. There is some controversy, however, concerning what exactly Bohr means by a "classical concept."[45] The interpretation that is endorsed here is that by "classical concepts" Bohr means simply the concepts of classical mechanics, such as "position," "momentum," "force," "electric field value," etc. Support for this interpretation can be found in quotations such as the following: "the unambiguous interpretation of any measurement must be essentially framed in terms of the

[44] Darrigol (1997) uses this phrase to describe the physics community's perception of Bohr's old quantum theory.

[45] Don Howard (1994) has offered what he calls a "reconstruction" of the notion of a classical concept that "seeks to be faithful to Bohr's words." On his view, by classical concepts "Bohr did not mean simply the application of classical physics – the physics of Newton, Maxwell"; he argues instead that by "classical" Bohr means "a description in terms of what physicists call 'mixtures'" (Howard 1994, p. 203).

classical physical theories, and we must say that in this sense the language of Newton and Maxwell will remain the language of physicists for all time" (Bohr 1931, p. 692; BCW 6, p. 360). Interpreting classical concepts in this way also coheres with Bohr's rational generalization thesis.

Bohr's claim that classical concepts must be used for an unambiguous communication of experimental results has been met with considerable incredulity and puzzlement – both from his contemporaries and from subsequent scholars. One prominent trend in current Bohr scholarship has been to try and make sense of this requirement in terms of a Kantian or neo-Kantian framework (a sample of such approaches can be found, for example, in Faye and Folse 1994). The approach adopted here, however, is to try and understand Bohr's doctrine of the indispensability of classical concepts in terms of his belief that quantum theory is a rational generalization of classical mechanics.

A point that has been overlooked in discussions of the doctrine of the indispensability of classical concepts is that there are really two distinct, though intertwined, ways in which Bohr takes these concepts to be indispensable. The first, and most often commented upon way, is Bohr's claim that the measuring instruments and the results of experiments must be expressed in terms of classical concepts. For example, Bohr writes,

> However far quantum effects transcend the scope of classical physical analysis, the account of the experimental arrangement and the record of the observations must always be expressed in common language supplemented with the terminology of classical physics. This is a simple logical demand, since the word "experiment" can in essence only be used in referring to a situation where we can tell others what we have done and what we have learned.
>
> *(Bohr 1948, p.313; BCW 7, p. 331)*

There is, however, a second and more subtle sense in which Bohr takes classical concepts to be indispensable. He expresses this second sense most clearly in his 1929 introductory essay:

> According to the view of the author, it would be a misconception to believe that the difficulties of the atomic theory may be evaded by eventually replacing the concepts of classical physics by new conceptual forms … No more is it likely that the fundamental concepts of the classical theories will ever become superfluous for the description of physical experience. The recognition of the indivisibility of the quantum of action, and the determination of its magnitude, not only depend on an analysis of measurements based on classical concepts, but it continues to be the application of these concepts alone that makes it possible to relate the symbolism of the quantum theory to the data of experience.
>
> *(Bohr [1929] 1934, p. 16; BCW 6, p. 294)*

In this quotation we see Bohr emphasizing that it is not only in the analysis of measurements that classical concepts are essential. These concepts are also essential for giving meaning to the abstract formalism of quantum theory; that is, they are necessary for connecting up this formalism with experience.[46]

A further clue to Bohr's view on the importance of classical mechanics for quantum theory is found in a little-discussed paper of Bohr's where he engages in a bit of counterfactual history. He asks us to consider a history of physics in which quantum mechanics had been discovered *before* classical mechanics:

> Imagine for a moment that the recent experimental discoveries of electron diffraction and photonic effects, which fall in so well with the quantum mechanical symbolism, were made before the work of Faraday and Maxwell. Of course, such a situation is unthinkable, since the interpretation of the experiments in question is essentially based on the concepts created by this work. But let us, nevertheless, take such a fanciful view and ask ourselves what the state of science would then be. I think it is not too much to say that we should be farther away from a consistent view of the properties of matter and light than Newton and Huygens were.
> *(Bohr 1931, p. 692; BCW 6, p. 360)*

In considering whether quantum mechanics could have been discovered first, Bohr immediately runs into the objection that this would be impossible, since the interpretation of the experiments that led to the discovery of quantum theory requires the use of classical concepts. This is the first sense of Bohr's doctrine of the indispensability of classical concepts discussed above. Bohr, however, sets this objection aside and pursues the thought experiment further to draw attention to the second sense in which classical concepts are indispensable. His conclusion is the following: quantum mechanics *by itself* provides a less adequate account of light and matter than does classical mechanics. This is a surprising conclusion to draw, especially given our current understanding of quantum mechanics as the more adequate theory that *replaced* classical mechanics. Bohr's point seems to be that quantum mechanics – without classical mechanics – is an inadequate theory. He is, of course, not saying that quantum theory is incomplete in the sense of the EPR debate, that is, that there is some element of reality that it leaves out of its description. Rather, it is incomplete in the sense that quantum mechanics depends on classical mechanics for its meaning – for connecting up its formalism with experience. Only by having classical and quantum mechanics together do we have an adequate physical theory.

4.5 Is Bohr a reductionist or pluralist?

When it comes to answering the question of what, according to Bohr, is the relation between classical and quantum mechanics, the usual options of reductionism (quantum mechanics reduces to classical mechanics in the appropriate limit) and

[46] For a further discussion of this point see Tanona (2004) and in connection with a discussion of "Bohr's measurement problem" see Bokulich and Bokulich (2005).

theoretical pluralism (each theory has its own proper domain of application) are inadequate. Explicating Bohr's view of the relationship between these two theories is complicated by the fact that there are elements of his view that can be identified with both the reductionist and the pluralist.

On the one hand, Bohr's view is reductionistic in the sense that he takes quantum mechanics to be a universal mechanical theory. This is made particularly clear in his debates with Erwin Schrödinger over the reality of stationary states. For example, he writes

> In the limit of large quantum numbers where the relative difference between adjacent stationary states vanishes asymptotically, [classical] mechanical pictures of electron motion may be rationally utilised. It must be emphasized, however, that this connexion cannot be regarded as a gradual transition towards the classical theory in the sense that the quantum postulate would lose its significance for high quantum numbers.
>
> *(Bohr 1928, p. 589; BCW 6, p. 157)*

This quotation also brings out another sense in which Bohr is often seen as a reductionist. Frequently in the context of his discussions of the correspondence principle, Bohr notes that the laws of the classical theory are suitable for the description of phenomena in a limiting region (e.g., Bohr 1924, p. 22; BCW 3, p. 479). This sounds a lot like reductionism in the sense that Nickles (1973) calls "reduction$_2$," which we examined in Chapter 1. In the above quotation, however, Bohr makes it clear that it is not the case that classical mechanics is recovered in any robust sense in this limit, rather it is simply that the classical algorithm provides an adequate approximation in this regime. In 1948 Bohr again emphasizes quantum theory's universal character:

> The construction and the functioning of all apparatus like diaphragms and shutters … will depend on properties of materials which are themselves essentially determined by the quantum of action. Still … we may to a very high degree of approximation disregard the molecular constitution of the measuring instruments.
>
> *(Bohr 1948, p. 315; BCW 7, p. 333)*

In this sense, Bohr – unlike Heisenberg – is not a theoretical pluralist; there is no regime for which classical mechanics is, strictly speaking, perfectly accurate or true.[47]

On the other hand, there is an aspect of Bohr's view of the relationship between classical and quantum mechanics that is more like theoretical pluralism than reductionism. Despite his assertion that quantum mechanics is a universal theory, Bohr is not an eliminativist – he does not think that classical mechanics, even in principle, can be eliminated. As we have seen in some detail, classical mechanics

[47] Recall the discussion of Heisenberg's closed theories in Chapter 2.

continues to play a very important role in physics, for Bohr, and it is not just for "engineering purposes."[48] On his view, quantum theory, without classical mechanics, is an inadequate – perhaps even meaningless – theory.

Through his rational generalization thesis, Bohr is offering us a new way of viewing the relationship between classical and quantum mechanics. Quantum mechanics is not a rival to classical mechanics, but rather a modification of it – a modification that depends on the applicability and consistency of the classical theory.

4.6 Conclusion

In the contemporary physics literature the correspondence principle is almost ubiquitously used to mean the requirement that the predictions of quantum mechanics match the predictions of classical mechanics in domains, such as $n \to \infty$, for which classical mechanics is empirically adequate. For example, Zurek and collaborators, in their article "Decoherence, chaos and the correspondence principle," define it as a demand for an agreement between the "quantum and classical expectation values" (Habib *et al.* 1998); or more generally Ford and Mantica, in their article "Does quantum mechanics obey the correspondence principle? Is it complete?", define the correspondence principle as requiring that "any two valid physical theories which have an overlap in their domains of validity must, to relevant accuracy, yield the same predictions for physical observations" (Ford and Mantica 1992, p. 1087). Although Bohr might have agreed with these requirements, it is certainly not what he meant by the correspondence principle.[49] Indeed when Bohr's student and collaborator Léon Rosenfeld suggested to Bohr that the correspondence principle was about the asymptotic agreement of quantum and classical predictions, Bohr emphatically protested and replied, "It is not the correspondence argument. The requirement that the quantum theory should go over to the classical description for low modes of frequency, is not at all a principle. It is an obvious requirement for the theory" (Rosenfeld [1973] 1979, p. 690).[50]

As we saw in Section 4.2, Bohr's correspondence principle is not some sort of asymptotic agreement of predictions, but rather what is better thought of as Bohr's selection rule: to each allowed quantum transition there is a corresponding harmonic in the classical periodic motion. In other words, despite the fundamental physical differences between the classical and quantum accounts of the atom and radiation,

[48] There is no one in the reductionism–pluralism debate who would deny the continued practical utility of classical mechanics for things like building bridges. This is not, however, the sense in which Bohr and the theoretical pluralists take classical mechanics to be indispensable.

[49] For a further criticism of these pervasive misinterpretations of Bohr's CP in the contemporary literature see also Batterman (1991).

[50] Quoted also in Tanona (2002, p. 7).

there is nonetheless a law-like connection between particular structures in the classical dynamics and the structure of the quantum dynamics. In light of Heisenberg's banishment of classical trajectories from matrix mechanics, it is all the more surprising that such a correspondence relation is preserved in the new quantum theory.

This proper understanding of the correspondence principle also helps us see that Bohr's account of the relation between classical and quantum mechanics is not simply a version of Nickels' reductionism$_2$ relation, with the relevant limit being $n \rightarrow \infty$.[51] Rather, Bohr viewed quantum mechanics as a rational generalization of classical mechanics; moreover, it is a generalization that depends essentially on the continued applicability of classical concepts. Like Dirac's view discussed in the last chapter, Bohr's account of the relation between classical and quantum mechanics does not fit neatly into the usual categories of either reductionism or pluralism, showing once again that there is a richer variety of ways of thinking about inter-theory relations than has usually been recognized.

A brief survey of references to Bohr's correspondence principle in the contemporary physics literature reveals that much of the interest in this (misunderstood) principle is due to the problem of quantum chaos. Indeed this is the subject of both Zurek's and Ford's articles mentioned above. As we saw in Chapter 1, despite the pervasiveness of chaotic behavior in classical systems, it is almost entirely absent in quantum systems. For chaotic systems, quantum mechanics can only mimic the classical behavior up to some limited time known as the "break time," after which the quantum and classical predictions diverge. While this is not the subject of Bohr's correspondence principle, it is a problem very relevant to Bohr's old quantum theory.

When one looks back over the history of the downfall of the old quantum theory, it turns out that many of the key problems plaguing this theory were due to the problem of quantum chaos. As we saw in Section 4.1, Einstein recognized early on the fundamental obstacle that nonintegrable systems pose for developing adequate quantization conditions. As Batterman (1991) in particular has emphasized, Bohr himself seems to have had some sense of this problem, as we see in the following quotation:

> For the purpose of fixing the stationary states, we have up to this point only considered simply or multiply periodic systems. However, as already mentioned in §1, the general solution of the [Hamilton] equations frequently yields motions of a far more complicated character. In such a case, the considerations previously discussed are not consistent with the existence and stability of stationary states ... But now, in order to give an account of the

[51] Recall the discussion of reductionism$_2$ given at the end of Section 1.2, and applied to the quantum–classical relation at the beginning of Section 1.4.

observed properties of the elements, we are forced to assume that the atoms … always possess "sharp" stationary states, although the general solution of the equations of motion for atoms with several electrons, even in the absence of external forces, exhibits no simple periodic properties of the kind mentioned.

(Bohr 1924, p. 15; BCW 3, p. 472)[52]

In this quotation, Bohr recognizes both that the typical classical motion for an atom with several electrons will not be periodic and that the entire apparatus of the old quantum theory, being based on the concept of stationary states, will not apply if the motion is non-periodic. As a further example, the failure of the old quantum theory to account for the ground and excited states of the helium atom can largely be attributed to the fact that the helium atom is a quantum system whose classical counterpart is the chaotic three-body problem. As we shall see in the next chapter, however, it turns out that, with a few modifications, the old quantum theory is in fact capable of solving the helium atom, and that far from being a theoretical dead end, the old quantum theory finds new life in the form of modern semiclassical mechanics.

[52] Batterman's (1991) insightful paper connecting Bohr's correspondence principle to modern semiclassical mechanics shall be discussed again in Section 7.1.

5

Semiclassical mechanics: Putting quantum flesh on classical bones

Though this be madness, yet there is method in't.

Shakespeare, Hamlet, *Act 2 Scene 2*

5.1 Introduction

Semiclassical mechanics can be broadly understood as the theoretical and experimental study of the interconnections between classical and quantum mechanics.[1] More narrowly, it is a field that uses classical quantities to investigate, calculate, and even explain quantum phenomena. Its methods involve an unorthodox blending of quantum and classical ideas, such as a classical trajectory with an associated quantum phase. For these reasons, semiclassical mechanics is often referred to as "putting quantum flesh on classical bones," where classical mechanics provides the skeletal framework on which quantum quantities are constructed.[2]

There are three primary motivations for semiclassical mechanics: First, in many systems of physical interest, a full quantum calculation is cumbersome or even unfeasible. Second, even when a full quantum calculation is within reach, semiclassical methods can often provide intuitive physical insight into a problem, when the quantum solutions are opaque. And, third, semiclassical investigations can lead to the discovery of new physical phenomena that have been overlooked by fully quantum-mechanical approaches. Semiclassical methods are ideally suited for studying physics in the so-called mesoscopic regime, which can roughly be

[1] The term "semiclassical mechanics" can be used even more broadly to designate all ray-theory approaches to wave theories, such as in optics or acoustics. Here I shall focus my discussion specifically on the semiclassical approach to quantum mechanics, though similar lessons could be drawn in these other fields. See Batterman (2002) for a discussion of some of these other contexts.

[2] This phrase is after Michael Berry and Karen Mount (1972), who titled a section of their paper on semiclassical mechanics "Sewing the wave flesh on the classical bones." According to Berry (personal communication), it is adapted from an expression used in the 1960s in the context of optics, which has been attributed originally to Boris Kinber.

understood as the domain between the classically described macro-world and the quantum mechanically described micro-world. An area in which semiclassical studies have proven to be particularly fruitful is in the subfield of quantum chaos. Although there can be no true chaos in the quantum realm (in the sense of quantum systems exhibiting sensitive dependence on initial conditions), surprising new phenomena do occur in those quantum systems whose classical counterparts are chaotic (Berry 1987; 1989). Semiclassical mechanics has played a central role in both the discovery and the explanation of these new "quantum chaos" phenomena.[3]

In Section 5.2, I introduce modern semiclassical mechanics, and show how it is related to both the old and new quantum theories. Within semiclassical mechanics there are two primary calculational tools: first, EBK quantization (named after Einstein, Brillouin, and Keller), which is applicable only to integrable systems, and second, periodic orbit theory (and its variant closed orbit theory), which is also applicable to systems that are classically chaotic. These periodic orbits are classical trajectories that close on themselves, and although they are very rare in chaotic systems, they form the backbone of semiclassical calculations. After reviewing the basics of modern semiclassical mechanics, I show in Section 5.3 how this theory was finally able to solve the helium atom semiclassically in the spirit of Bohr's old quantum theory. Even though the helium atom has long been "solved" by modern quantum mechanics, a semiclassical solution of this atom is particularly valuable insofar as it is able to provide new physical insight into the dynamical structure of this quantum system in a way that the purely quantum-mechanical solutions do not.

Not only have classical trajectories proven to be a reliable tool for calculating quantum quantities, but even more remarkably, recent experiments in quantum chaos have revealed new quantum phenomena that seem to be shaped by these classical periodic orbits. In other words, these classical dynamical structures, which were thought to be mere calculational devices, are manifesting themselves in surprising ways in quantum experiments. In the remaining sections of this chapter I discuss in detail two examples of such experiments. The first case example, presented in Section 5.4, involves measuring the spectra of highly excited atoms, such as hydrogen, in a strong magnetic field (a quantum system whose classical counterpart is chaotic). These experiments reveal unexpected oscillations in the quantum spectrum that seem to be explained by particular classical closed trajectories, namely those trajectories that the excited electron of an atom in a magnetic field *would* follow *if* classical mechanics were true. Even more dramatically, subsequent experiments have shown that the detailed properties of these fictitious classical trajectories can be measured directly from the experimental quantum

[3] The explanatory importance of semiclassical mechanics was first appreciated in the philosophical literature by Batterman (1993; 1995; 2002), and will be defended more fully in Chapter 6.

data. This has led some researchers to question whether these classical trajectories should really be thought of as fictitious.

The second case example, presented in Section 5.5, comes from numerical experiments performed on systems known as quantum billiards, which are the quantum analog of classically chaotic billiard models.[4] These highly accurate simulations have revealed that the quantum wavefunctions of these systems can exhibit an unexpected accumulation of probability density along the trajectories of the rare periodic orbits of the classical billiard system.[5] This quantum phenomenon, known as wavefunction scarring, presents another striking example of the way in which classical dynamical structures are manifesting themselves in quantum phenomena. Just as in the case of the anomalous oscillations in the quantum spectra, one finds that the explanation of this quantum phenomenon makes central appeal to these "fictitious" classical trajectories.[6]

In the last section I conclude that there are three lessons that we should take away from this research in semiclassical mechanics: First, the utility of using classical concepts, such as trajectories, in quantum phenomena did not in fact end with the downfall of the old quantum theory. There is a domain of quantum systems for which semiclassical mechanics, with its blending of classical and quantum ideas, provides the appropriate theoretical framework for investigating, calculating, and explaining these quantum phenomena. Second, these experiments have revealed that there is a continuity of dynamical structure between classical and quantum mechanics, which underpins these semiclassical methods. And finally, this research in semiclassical mechanics, and especially in the subfield of quantum chaos, has revealed that the relationship between classical and quantum mechanics is much more subtle and intricate than the simple statement $\hbar \to 0$ might lead us believe.

5.2 A phoenix from the ashes: Semiclassical mechanics

Almost as soon as the new quantum theory was introduced, the need for a method to obtain approximate solutions to the Schrödinger equation was recognized, since analytic solutions to the Schrödinger equation can only be obtained for a few simple systems. In 1926, Gregor Wentzel, Hendrik Kramers, and Leon Brillouin independently introduced an approximation method, now known as the WKB method (or "phase integral method"), that is asymptotic to the full quantum solution as $\hbar \to 0$. What is typically not emphasized in current textbook treatments of WKB, however, is the close relation between this method of approximation and the quantum conditions of the old quantum theory. Both Kramers and Brillouin explicitly show

[4] For a defense of the claim that numerical simulations can be properly thought of as a kind of experiment, see Parker (forthcoming).

[5] These simulations have since been confirmed experimentally in the lab by Sridhar (1991), and the same scanning phenomenon was observed.

[6] The ontological status of these trajectories shall be examined in more detail in Chapter 6.

in their papers how the Bohr–Sommerfeld quantum conditions can be derived from their approximations. While Brillouin's derivation leads to the standard quantum conditions, Kramers' approximation method, with its "connection formula" relating the approximate wavefunctions at the classical turning points, resulted in a modified quantum condition with half-integer quantum numbers:

$$J = \oint p\mathrm{d}q = (n + 1/2)h. \tag{5.1}$$

Kramers concludes from this that the debate over integer versus half-integer quantum conditions, which had been cut short by the invention of the new quantum theory, had now been solved in favor of half-integers (Kramers 1926, p. 833). Early on, the WKB method was seen not just as a useful approximation tool, but as a technique that leads to new and improved insight into the old quantum theory. For example, Young and Uhlenbeck write of the WKB method,

> The value of this approximate solution is two-fold; first, of course, it can lead to valuable results in cases where the exact solution is unknown or difficult to handle; in the second place it has the advantage that it is a link between the classical quantum theory of Bohr and the exact quantum mechanics. One may say that it shows how from the viewpoint of the latter, at least in the case of one-dimensional problems, the classical physicist ought to have treated his problems to obtain the best results. We illustrate this in general by showing that the approximate solution leads to a determination of the energies by quantization of the classical action integrals with the sole difference that half-integer quantum numbers must be used instead of whole numbers.
>
> *(Young and Uhlenbeck 1930, p. 1156)*

In other words, the WKB approximation can also be understood as leading to a further refinement of the Bohr–Sommerfeld quantum conditions, one that places the old quantum theory on a firmer foundation as an approximation to the full quantum mechanics.

A limitation of the standard WKB method is that it can only be applied to one-dimensional problems, and so to systems for which the Schrödinger equation is separable.[7] In the 1950s the applied mathematician Joseph Keller added yet another chapter to the history of the quantum conditions with his paper titled "Corrected Bohr-Sommerfeld quantum conditions for nonseparable systems." In this paper, Keller shows how Einstein's 1917 generalization, which we discussed in Section 4.1, can be

[7] The following clarification of terminology may be helpful. A Hamiltonian system is said to be *integrable* if there are as many integrals (constants) of the motion as there are degrees of freedom (n), and those integrals are in involution, meaning that they all commute with each other. This means that for an integrable system, the phase space trajectory will be confined to an n-dimensional manifold which has the topology of a torus (recall the discussion of Einstein's "rationalized coordinate space" in Section 4.1, and for further explanation see Tabor (1989), Chapter 2). A system is said to be *separable* if in addition to being integrable, those integrals of the motion are also single-valued (Landau and Lifshitz 1981, p. 166), meaning each of them is a function of one coordinate only. Hence, all separable systems are integrable, but not all integrable systems are separable. Nonetheless, nonseparable integrable systems are rare, so these terms are sometimes used as roughly interchangeable.

combined with Brillouin's WKB derivation to develop a more adequate formulation of the quantum conditions that can be applied to nonseparable systems as well. Keller begins by writing an approximation to the wavefunction as a sum of waves:

$$\psi(q) \sim \sum_{k=1}^{M} A_k(q) \mathrm{e}^{\mathrm{i}(h/2\pi)S_k(q)}, \tag{5.2}$$

where A is the amplitude and S is the phase, which is interpreted as the classical action function.[8] The fact that quantum mechanics requires the wavefunction ψ to be a single-valued function of the coordinates imposes the following constraint on each of the S_k and A_k :

$$\Delta S = \left[n + \mathrm{i} \frac{\Delta \log A}{2\pi} \right] h. \tag{5.3}$$

In coordinate space, the differences ΔS and ΔA can be expressed as line integrals over closed curves in coordinate space, yielding

$$\oint \nabla S \cdot \mathrm{d}s = \left[n + \frac{\mathrm{i}}{2\pi} \oint \nabla \log A \cdot \mathrm{d}s \right] h. \tag{5.4}$$

Using $\nabla S \cdot \mathrm{d}s = \sum_i p_i \mathrm{d}q_i$, Keller obtains

$$\oint \sum_i p_i \mathrm{d}q_i = \left[n + \frac{\mathrm{i}}{2\pi} \oint \nabla \log A \cdot \mathrm{d}s \right] h, \tag{5.5}$$

which, in the special case where A is single valued, reduces to Einstein's quantum condition (Equation 4.5). In order to evaluate the change in log A around a closed curve, one needs to determine the number of times m, that the curve passes through a caustic, which is the higher-dimensional analog of a turning point.[9] Just as in Kramer's WKB derivation there are connection formulas, which show that every time a trajectory touches a caustic, the phase of A changes by $-\pi/2$. Using this information one obtains

$$\frac{\mathrm{i}}{2\pi} \oint \nabla \log A \cdot \mathrm{d}s = \frac{\mathrm{i}}{2\pi} m \left(-\frac{\mathrm{i}\pi}{2} \right) = \frac{m}{4}, \tag{5.6}$$

which when substituted into Equation 5.5 above yields Keller's "corrected Bohr–Sommerfeld" quantum conditions:

[8] The number of terms, M, depends on the particular potential. For a more detailed account of Keller's derivation see Keller (1958; 1985), which are summarized here.

[9] "m" is also known as Maslov's index, after the Russian mathematician Viktor Maslov who discovered it independently of Keller.

$$\oint \sum_{i=1}^{N} p_i dq_i = \left(n + \frac{m}{4}\right) h. \tag{5.7}$$

These quantum conditions are known more generally as the EBK quantization rules after Einstein, Brillouin, and Keller, or "torus quantization" due to the geometrical interpretation of this formula.[10] Keller summarizes this approach as follows:

> The method we have described is often called semiclassical mechanics. It employs various elements of classical mechanics: the trajectories, the action function, the Hamilton-Jacobi equation, the Liouville equation, Hamilton's equations, etc. They are used to construct an asymptotic solution of the Schrödinger equation, which is sometimes called the semiclassical wave function, and to obtain the corrected quantum conditions. These conditions in turn yield asymptotic approximations to the energy eigenvalues of the Schrödinger equation.
>
> *(Keller 1985, p. 494)*

Although Keller's quantum conditions are an improvement over Einstein's in so far as they take into proper account the role of caustics, they do not, as Keller recognizes, overcome Einstein's more fundamental concern, namely that they only work if the system is integrable.[11]

It was not until the 1970s that a semiclassical method applicable to classically chaotic systems was devised by Martin Gutzwiller. Although Gutzwiller's semiclassical method is importantly different from the WKB and EBK methods for integrable systems, he nonetheless sees it as being an extension of the quantum conditions of the old quantum theory. He writes, "the phase-integral approximation is extended to systems which are not multiply periodic, and quantization conditions à la Bohr and Sommerfeld are derived for cases in which none had been formulated so far" (Gutzwiller 1971, p. 343). The centerpiece of Gutzwiller's method is the trace formula, which allows one to calculate the quantum density of states from the classical periodic orbits (that is, those classical trajectories that close on themselves). Even in a fully chaotic system there is an infinite number of periodic orbits, though they are isolated, unstable, and are of measure zero.

For a bound system with Hamiltonian $\hat{H}$, the spectrum is determined by $\hat{H}|n\rangle = E_n|\, n\,\rangle$, where the E_n are the discrete energy levels of the system and $|n\rangle$ are the eigenstates; the density of states is then defined as $\rho(E) = \mathrm{Tr}\,[\delta(E - \hat{H})]$, which can be evaluated as $\rho(E) = \sum_n \langle n|\delta(E - \hat{H})|n\rangle = \sum_n \delta(E - E_n)$. The density of states can be expressed as the sum of two terms: $\rho(E) = \bar{\rho}(E) + \rho_{\mathrm{osc}}(E)$, which are the familiar mean level density plus an oscillatory term. The Gutzwiller trace formula

[10] For an excellent and detailed introduction to torus quantization see Tabor (1989), Chapter 6.

[11] Keller notes "On the negative side is the demonstration of the widespread occurrence of chaotic solutions by numerical computation and by theoretical analysis. New methods must be devised to apply semiclassical mechanics to such solutions (Keller 1985, p. 485).

shows how this quantity can be approximated by a sum over all the classical periodic orbits of the system:

$$\rho_{\mathrm{osc}}(E) \approx \frac{1}{\pi\hbar}\sum_{p} A_p \cos\left(\frac{S_p}{\hbar} - m_p\frac{\pi}{2}\right), \tag{5.8}$$

where p labels the periodic orbits, S is the action of those periodic orbits, A is determined by the period and stability of the orbits, and m is an index that, similar to the m in the EBK formula, counts the number of caustics or conjugate points encountered by the trajectory.[12]

What makes Gutzwiller's trace formula a semiclassical formula is that it shows us how to determine a quantum quantity (the density of states) in terms of classical quantities (the classical periodic trajectories and their properties). Although the sum does not converge, this problem can often be overcome by using so-called cycle expansion techniques, or even by considering only a handful of periodic orbits, which has surprisingly been shown to often yield remarkably accurate results.[13] As we shall see in the next section, it was these very techniques that finally led to a successful semiclassical solution of the helium atom.

5.3 The helium atom solved

As we saw in the beginning of Chapter 4, a catalyst for the rejection of Bohr's old quantum theory was the inability of the theory in the early 1920s to obtain a reasonable estimate of either the ground or excited states of the helium atom. Fifty-five years after the introduction of the new quantum theory, J. Leopold and Ian Percival reexamined the helium atom using modern semiclassical methods and made the following surprising declaration: "the failure of the old quantum theory to obtain an accurate energy [for the ground state of helium] is … *not* [due] to any intrinsic fault of the theory" (Leopold and Percival 1980, p. 1037). Instead they argue that the failure was due to an ignorance of the importance of Maslov indices, which give rise to half-integer quantum numbers. Leopold and Percival used the EBK quantization method to obtain a value for the ground state of helium that is comparable to the "successful crucial experiment" value initially obtained by Hylleraas in 1929.[14] Although the inclusion of Maslov indices brings the semiclassical approach of the old quantum theory into closer agreement with experiment, by contemporary standards, their results are still inadequate.

[12] For a more detailed account of Gutzwiller's trace formula see Gutzwiller (1990).

[13] The cycle expansion techniques have been developed by Cvitanović and colleagues, and these techniques were central to the semiclassical solution of the helium atom discussed below (see Cvitanović *et al.* (2005) for further details). Regarding the success of using only a small number of periodic orbits in the sum in the context of solving the helium atom, see, for example, Ezra *et al.* (1991) and Wintgen *et al.* (1992).

[14] Recall the discussion of Hylleraas at the end of Section 4.1.

The limited success of Leopold and Percival's results is perhaps not surprising in light of Einstein's warnings in 1917 that the torus (EBK) quantization conditions cannot be applied to nonintegrable systems such as the three-body problem. The helium atom is a nonintegrable system with a mixed phase space, and it was only quite recently demonstrated that a considerable part of it is chaotic (Wintgen *et al.* 1992). For example, if one considers the collinear configuration of the helium atom, in which the positions of the nucleus and two electrons lie along a line, then the case where the two electrons are on the same side of the nucleus turns out to be regular, while the configuration with one electron on each side of the nucleus (a configuration that is favored since electron–electron interaction is minimized) is fully chaotic (Wintgen *et al.* 1992). Because the helium atom can exhibit chaotic behavior, one must use Gutzwiller's semiclassical method.

A proper semiclassical solution of the helium atom was finally achieved by Dieter Wintgen and collaborators in 1991 (Ezra *et al.* 1991; see also Wintgen *et al.* 1992). Their work demonstrated that an accurate calculation of the ground and excited states of the helium atom is in fact possible using a semiclassical approach of the sort initiated by Bohr. They showed that one can accurately calculate the helium spectrum using the fictional assumption that electrons orbit the nucleus on classical "planetary" trajectories.

One of the leading researchers in the field, Predrag Cvitanović, remarkably describes the failure of the old quantum theory and the success of Wintgen *et al.* as follows:

> Why did Pauli and Heisenberg fail with the helium atom? It was not the fault of the old quantum mechanics, but rather it reflected their lack of understanding of the subtleties of classical mechanics. Today we know what it is that they missed in 1913–24: the role of conjugate points (Maslov indices) along classical trajectories was not accounted for, and they had no clue of what the role of periodic orbits in nonintegrable systems should be. Since then the calculation for helium using the methods of the old quantum mechanics has been fixed. Leopold and Percival added the Maslov indices in 1980, and in 1991 Wintgen and collaborators understood the role of periodic orbits ... One is also free to ponder what would quantum theory look like today if all this was worked out in 1917. When asked this question, Hans Bethe responded with an exasperated look: it would be just the same, he said.
>
> *(Cvitanović* et al. *2005, p. 3)*

The view that Cvitanović expresses here is one shared by many researchers in the field: the old quantum theory was not a theoretical dead end – and even more interestingly, what was required to turn this research program around, was not some further quantum insight, but rather a deeper understanding of classical mechanics. The success of a semiclassical approach depends on an adequate understanding of the classical mechanical problem, and one of the first things that Wintgen and collaborators had to do to solve the *quantum* helium atom, was to develop new results for the *classical* three-body Coulomb problem.[15] It is important to emphasize that these researchers are not claiming that the old quantum theory is the true theory, nor that modern quantum mechanics should be abandoned. Rather, the claim is simply that the old quantum theory has not yet exhausted its usefulness, and, with further development, it has the resources to continue to be a fruitful theory, leading to new insights and discoveries.[16]

The interest in the semiclassical solution of the helium atom is not simply to resurrect the old quantum theory by solving one of its most famous problems. Rather, the value of this approach lies in the ability of semiclassical mechanics to provide calculational and physical insight into situations where full quantum solutions are either unavailable or opaque. In his paper providing the first semiclassical solution of the helium atom, Wintgen explains,

> A semiclassical description of two-electron atoms is also highly desirable, because most parts of the spectral regions are still unexplored, both experimentally and quantum theoretically … The high dimensionality of the problem combined with the vast density of states can make the [full quantum mechanical] calculations cumbersome and elaborate … Furthermore, the problem of understanding the structure of the quantum solutions still remains after solving the Schrödinger equation. Again, simple interpretation of classical and semiclassical methods assists in illuminating the structure of solutions.
>
> (*Wintgen* et al. *1992, p. 19)*

In this quotation we see three reasons being offered for why a semiclassical treatment of atoms such as helium is desirable. First, the semiclassical treatments provide an *investigative tool*: semiclassical methods allow one to investigate physical domains that might not yet be accessible either experimentally or with a fully quantum approach. Second, they provide a *calculational tool*: semiclassical calculations, though far from trivial, can be less cumbersome than full quantum calculations. Finally, they provide an *interpretive tool*: semiclassical methods can offer physical insight into the structure of a problem, in a way that a fully quantum-mechanical approach might not.

While few would question the usefulness of semiclassical mechanics as a calculational tool, it perhaps comes as more of a surprise that semiclassical mechanics is also an important interpretive tool, offering physical insight into situations in a way that the fundamental theory does not. As an example of the sort of structural insight that semiclassical approaches can offer, Wintgen *et al.* discuss how the semiclassical calculation of the quantum-mechanical density of states for the helium atom reveals that a long-held assumption about the quantum behavior of the electrons in this atom is in fact mistaken. They write,

[15] As I shall argue in Chapter 7, this sort of methodology is quite similar to Dirac's "reciprocal correspondence principle" method described in Section 3.2.

[16] Examples of such new insights and discoveries yielded by semiclassical approaches shall be given below.

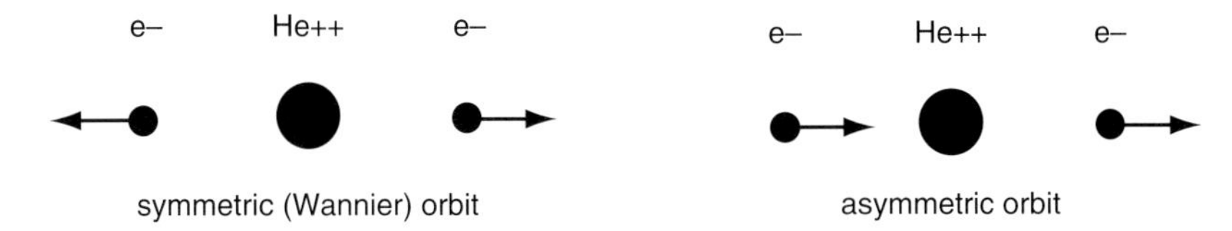

Figure 5.1 The symmetric and asymmetric stretch orbits of the collinear helium atom.

> The interpretive ability of the [semiclassical] methods illuminates the structure of the quantal motion. The analysis shows that the near-collinear intra-shell resonances are associated with the (fundamental) asymmetric stretch like motion of the electron pair. Semiclassically, this observation is nearly trivial. The result is nevertheless remarkable, in that it has been widely believed for decades that these resonances are associated with the in-phase symmetric stretch motion of the electron pair along the Wannier ridge.
>
> *(Wintgen* et al. *1992, p. 32)*

Intra-shell resonances are excited states of helium, in which both electrons are highly excited in a symmetric fashion. The symmetry of the two excited electrons means that there are strong electron–electron correlations, rendering the familiar Hartree–Fock methods inapplicable. It had long been believed that the wavefunctions of these resonant states were localized along the so-called Wannier ridge (also known as the symmetric stretch orbit), in which the electrons move in phase in Kepler ellipses (or in the strict collinear case, back and forth along a single line) in such a way that the distances of the electrons from the nucleus is always the same. This is contrasted with the asymmetric stretch orbit in which the distances of the two electrons from nucleus are not equal (see Figure 5.1).

It turns out that the symmetric (Wannier) orbit is highly unstable, with a divergent Lyapunov (stability) exponent. As a result, it cannot give rise to resonant structures in the density of states (Wintgen *et al.* 1992, p. 23). The asymmetric stretch orbit, on the other hand, is only weakly unstable. When this asymmetric orbit is semiclassically quantized, it yields surprisingly accurate results for the intrashell resonant states. This result indicates that the resonances are associated with this orbit, and not the symmetric orbit as had been believed. This conclusion has since been confirmed by accurate solutions of the full Schrödinger equation for the intrashell wavefunctions (Ezra *et al.* 1991, p. L418). As Wintgen *et al.* note, however, it is not just that the semiclassical approach yields satisfactory values for the energies of the helium atom, but "even more valuable [is] that the (semi-) classical analysis provides an insight of what the electrons are actually 'doing' in the highly correlated states" (Wintgen *et al.* 1992, p. 28). That is, the semiclassical analysis provides physical insight into the dynamical structure of the quantum system in a way that the full quantum calculations, though correct, do not.

5.4 Rydberg atoms and electron trajectories

Another area of research in which semiclassical approaches are yielding important structural insights is, ironically, another area in which the old quantum theory initially faltered: the spectra of atoms in external electromagnetic fields. The Zeeman effect is the splitting of spectral lines when an atom is placed in a magnetic field. While classical mechanics and the old quantum theory could explain the normal Zeeman effect, in which a single spectral line is split into a triplet, it could not explain more complex patterns of splitting, which were referred to as the "anomalous" Zeeman effect. For many of the founders of quantum theory, the inability of the old quantum theory to explain these spectral anomalies played a central role in the banishment of electron trajectories from quantum systems. Wolfgang Pauli, for example, writes in 1925,

> How deep the failure of known theoretical principles is, appears most clearly in the multiplet structure of spectra ... One cannot do justice to the simplicity of these regularities within the framework of the usual principles of the quantum theory. It even seems that one must renounce the practice of attributing to the electrons in the stationary states trajectories that are uniquely defined in the sense of ordinary kinematics.
>
> *(Pauli 1926, p. 167; quoted in Darrigol 1992, pp. 181–2)*

In Pauli's view, the notion that an electron is following a definite trajectory is incompatible with the spectroscopic data from atoms in magnetic fields. The mystery of the anomalous Zeeman effect was ultimately explained by the introduction of a fourth degree of freedom for the electron, which would later be known as spin.[17] Somewhat surprisingly, however, even after some eighty years of modern quantum mechanics, a full understanding of the behavior of atoms in magnetic fields is still an outstanding problem.

The spectroscopic data and Zeeman effects that most of us are familiar with (including the anomalous and Paschen–Bach effects) take place in the regime in which the external magnetic field is relatively weak compared with the electrostatic Coulomb field of the atom.[18] If, however, the magnetic field strength is increased so that it is comparable to the Coulomb field, then a diversity of new phenomena occur, collectively known as the quadratic Zeeman effect.[19] This is true even for the simplest of atoms, namely the hydrogen atom. While the hydrogen atom in a weak field is

[17] The discovery of spin is attributed to Ralph Kronig, George Uhlenbeck, and Samuel Goudsmit. For a nice history of this discovery and Pauli's role in solving the anomalous Zeeman effect, see Massimi (2005).

[18] See Delos *et al.* (1983) for a discussion of the different magnetic field strength regimes.

[19] The quadratic Zeeman effect is so named because the Hamiltonian of an atom such as hydrogen in a magnetic field has two terms involving the magnetic field, B: one that is linear in B and one quadratic in B (that is, it has B^2). For a sufficiently weak magnetic field, one can ignore the quadratic term in the Hamiltonian and only the linear term is important; if however the magnetic field is very strong, then the quadratic term cannot be neglected. Atoms in strong magnetic fields are often referred to as diamagnetic atoms.

nearly integrable, the dynamics of a hydrogen atom in a very strong magnetic field is nonintegrable and classically chaotic.[20] Hence these systems promise new insight into quantum chaos, that is, the way that classical chaos manifests itself in quantum mechanics.

Of particular interest to semiclassical mechanics, are so-called Rydberg atoms. These atoms, which were named after the late-nineteenth-century Swedish spectroscopist Johannes Rydberg, are atoms in which the outermost electron has been excited to a very high energy level.[21] Because the outer electron is in such a highly excited state, the dimension of these atoms is enormous, approaching the size of a fine grain of sand. These atoms call to mind Tom Stoppard's play *Hapgood*, in which he writes "There is a straight ladder from the atom to the grain of sand, and the only real mystery in physics is the missing rung. Below it, [quantum] particle physics; above it, classical physics; but in between, metaphysics" (Stoppard 1988, Act I Scene 5).[22] As an atom that *is* the size of a grain of sand, Rydberg atoms are ideal tools for studying the "metaphysics" of the relationship between classical and quantum mechanics.

Interest in the behavior of Rydberg atoms in strong magnetic fields (also referred to as diamagnetic Rydberg atoms) comes not just from theoretical questions about the relation between classical and quantum mechanics, but from a new generation of "anomalous" experimental spectroscopic data. In a series of experiments beginning in 1969 William Garton and Frank Tomkins at the Argonne National Laboratory examined the spectra of highly excited barium atoms in a strong magnetic field (Garton and Tomkins 1969). When the magnetic field was off, these Rydberg atoms behaved as expected: as the energy of the photons being used to excite the atom increased, there were a series of peaks at the energies which the barium atom could absorb the photons; and when the ionization energy was reached (that is, the energy at which the outer electron is torn off, leaving a positive ion), there were no more peaks in the absorption spectrum, corresponding to the fact that the barium atom could no longer absorb any photons. However, when they applied a strong magnetic field to these barium atoms and repeated this procedure, a surprising phenomenon occurred: the barium atoms continued to yield absorption peaks long after the ionization energy had been reached and passed (see Figure 5.2).

[20] Whether or not the hydrogen atom is chaotic depends on the strength of the magnetic field and on the value of angular momentum (Robnik 1981).
[21] Although Rydberg atoms are typically created in the lab by photoexciting an atom with laser light, Rydberg atoms do occur in nature, being relatively common in interstellar space.
[22] von Baeyer (1995) in a nice popular science article also refers to this quotation in connection with research on Rydberg atoms.

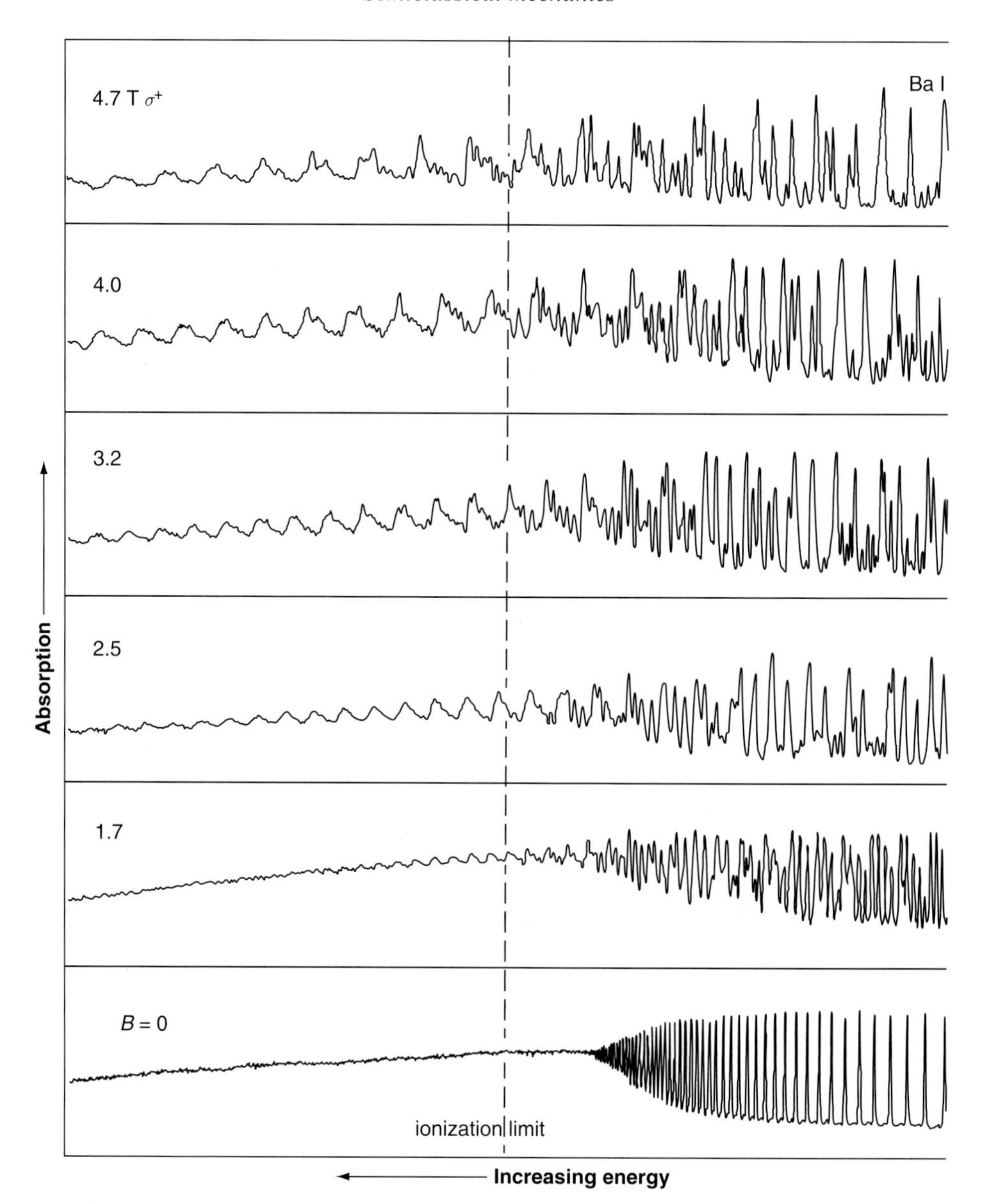

Figure 5.2 The absorption spectrum of barium. Higher energies are to the left, and the vertical dashed line is the ionization threshold, energies above which the barium atom ionizes. The bottom row is the spectrum with no magnetic field ($B = 0$), and the subsequent rows above that are the spectra with a magnetic field at 1.7, 2.5, 3.2, 4.0, 4.7 Tesla, respectively. Note the surprising oscillations in the absorption spectrum above the ionization limit when a strong magnetic field is present. (Adapted from Lu *et al.* 1978, Figure 1).

These oscillations in the spectrum were later named "quasi-Landau" resonances, and were shown to have a spacing independent of the particular type of atom. Remarkably, even almost twenty years after these quasi-Landau resonances were first discovered, a full theoretical explanation of them remained an outstanding problem.

The situation was further exacerbated by the fact that experimentalists were continuing to find new resonances above the ionization limit. For example, higher resolution experiments on a hydrogen atom in a strong magnetic field, performed by Karl Welge's group in Bielefeld in the mid-1980s, revealed many more types of resonances in the absorption spectrum (Main *et al.* 1986; Holle *et al.* 1988).[23]

Furthermore, these new resonances seemed to have lost the regularity of the quasi-Landau resonances discovered earlier. Instead, this new high-resolution spectral data exhibited a complex irregular pattern of lines. By the end of the 1980s, the MIT experimentalist Daniel Kleppner and his colleagues write,

> A Rydberg atom in a strong magnetic field challenges quantum mechanics because it is one of the simplest experimentally realizable systems for which there is no general solution ... We believe that an explanation of these long-lived resonances poses a critical test for atomic theory and must be part of any comprehensive explanation of the connection between quantum mechanics and classical chaos.
>
> (*Welch* et al. *1989, p. 1975*)

Once again the Zeeman effect was yielding anomalous spectra, whose explanation seemed to require the development of a new theoretical framework. Although modern quantum mechanics was never in doubt, the mesoscopic nature of Rydberg atoms suggested that an adequate theoretical explanation of these resonance phenomena would require not only quantum mechanics, but concepts from classical chaos as well.

An important step towards explaining these resonances was made by the Bielefeld group in a subsequent paper. They realized that if one takes the Fourier transform of the complex and irregular looking spectra, an orderly set of strong peaks emerges in the time domain (see Figure 5.3).

This resulting "recurrence spectrum" revealed that the positions of these peaks in the time domain were precisely at the periods, or transit times, of the classically allowed closed orbits for the electron moving in the combined Coulomb and magnetic fields; that is, each peak in the quantum spectrum corresponds to a different closed classical trajectory. They write,

> Though those experiments [by Welge's group in the 1980s] suggested the existence of even more resonances their structure and significance remained fully obscure. In this work we

[23] Even these higher resolution experiments were still of finite resolution, not resolving individual energy levels. What Figure (5.3) below shows is the average absorption curve as a function of energy.

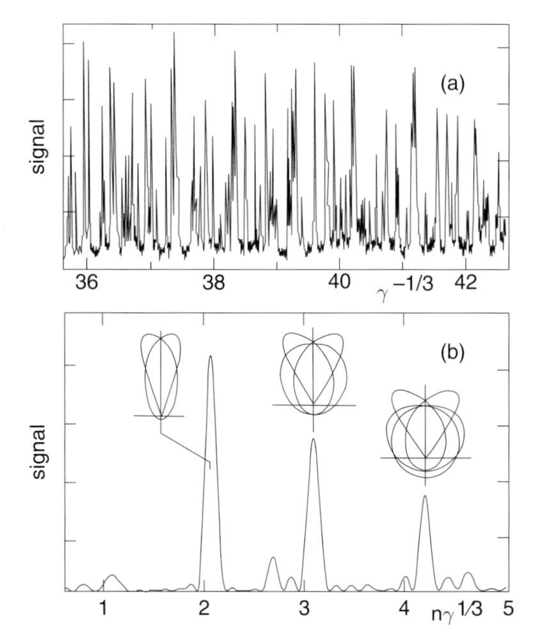

Figure 5.3 (a) The irregular looking scaled energy spectrum for a hydrogen atom near the ionization limit in a strong magnetic field. (b) The Fourier transform of this same spectrum into the time domain. The particular classical closed orbit corresponding to each of these well-defined peaks is superimposed. (Holle *et al.* 1988, Figure 1; courtesy of J. Main).

have discovered the resonances to form a series of strikingly simple and regular organization, not previously anticipated or predicted … The regular type resonances can be physically rationalized and explained by *classical* periodic orbits of the electron on closed trajectories starting at and returning to the proton as origin with an orbital recurrence-time T characteristic for each ν-type resonance.

(Main et al. *1986, pp. 2789–90)*[24]

Their results revealed the surprising ways in which the electron trajectories of classical mechanics are manifesting themselves in the experimentally obtained quantum spectrum. In other words, the explanation of these anomalous resonances and their regular organization seems to be intimately tied to the fictional assumption that these Rydberg electrons, instead of behaving quantum mechanically, are following definite classical trajectories.

[24] 'ν-type' just refers to the clearly observable strong peaks in the Fourier-transformed spectra, with each peak being labeled with an integer, ν.

Initial attempts to explain these oscillations in the spectrum tried to make use of the EBK semiclassical approach (discussed in Section 5.2).[25] However, near the ionization energy, the typical electron trajectory for a hydrogen atom in a strong magnetic field is chaotic; there is no smooth vector field $p(q)$ and the system is nonintegrable. As Einstein had recognized back in 1917, such a semiclassical approach cannot be used for these sort of systems.

A full quantitative semiclassical framework capable of explaining these anomalous resonances was finally introduced two years later by John Delos and his student Meng-Li Du (Delos and Du 1988; Du and Delos 1988). The close connection between each of the positions of these peaks in the spectrum and the transit times of the electron on a particular closed classical trajectory, suggested that a semiclassical approach, such as Gutzwiller's periodic orbit theory, would be the key to explaining the experimental data. In keeping with the anomalous experimental data to be explained, Delos and Du's semiclassical theory is designed to describe the quantum spectrum at a finite resolution; that is, a resolution in which the individual energy levels are not resolved. Their theory, which is known as "closed orbit theory," is similar to Gutzwiller's in that it is grounded in a semiclassical approximation to the Green's function, though instead of using periodic orbits, Delos's theory makes use of closed orbits, namely, those orbits that are launched from, and return to, the vicinity of the nucleus.

The Green's function $G_E^+(\vec{q};\vec{q}\,')$, recall, is the wave of energy E that is produced at $\vec{q}$ from a source at $\vec{q}\,'$. One can then construct a semiclassical approximation to the time-independent Green's function as follows:[26]

$$G_E^+(\vec{q};\vec{q}\,') = c\sum_k |\rho_k(\vec{q};\vec{q}\,')|^{1/2}\cdot\exp[\mathrm{i}S_k(\vec{q};\vec{q}\,')/\hbar - \mathrm{i}\mu_k\pi/2], \tag{5.9}$$

where the sum is over all trajectories from an initial point $\vec{q}\,'$ to a final point $\vec{q}$, S is the classical action function, ρ is the classical density, and μ is the Maslov index counting the caustics. It is a semiclassical approximation in so far as each term in the above sum approximately satisfies the Schrödinger equation as $\hbar \to 0$ (Delos and Du 1988, p. 1449). One can then calculate the density of states from the Green's function via the relation

$$\rho(E) = -(1/\pi)\int \mathrm{Im}G_E^+(\vec{q};\vec{q}\,')\cdot\delta(\vec{q}-\vec{q}\,')\mathrm{d}\vec{q}\mathrm{d}\vec{q}\,'. \tag{5.10}$$

[25] See references given in Du and Delos 1988, p. 1899.
[26] The presentation here follows Delos and Du 1988 and Du and Delos 1988, where further details can be found.

The delta function means that the final point is the same as the source point, and furthermore, when the above integral is evaluated according to the stationary phase method, the only semiclassical terms that will contribute are those for which the initial and final momenta are equal – that is, the periodic orbits. One can then obtain the following approximate formula for the density of states:

$$\rho(E) = \rho_0(E) + \sum_k a_k(E) \sin[S_k(E)/\hbar + \eta_k], \tag{5.11}$$

where the $\rho_0(E)$ is the contribution to the density of states from the singular term in $G_E^+(\vec{q};\vec{q}\,')$, and each of the other terms is the result of one of the periodic orbits, with η_k keeping track of the phases (such as the Maslov index), and a_k being an amplitude.[27] Using the relation $T_k = \partial S_k(\vec{q};\vec{q}\,')/\partial E$ one can rewrite Equation (5.11) as

$$\rho(E) = \rho_0(E) + \sum_k a_k(E) \sin\left[\int_{E_0}^{E} T_k(E')\mathrm{d}E'/\hbar + \eta_k\right], \tag{5.12}$$

which over a small energy range can be approximated as

$$\rho(E) \approx \rho_0(E) + \sum_k a_k \sin(T_k E/\hbar + \eta_k), \tag{5.13}$$

where $T_k(E)$ is the time of the particle to complete one period on the kth orbit of energy E. It is interesting to note that Delos and Du consider this result to be a new kind of correspondence principle. They write,

> We have therefore arrived at what could be called a correspondence principle for finite-resolution spectra. The principle is expressed quantitatively by [Equations (5.11)–(5.13)], or qualitatively by the following statement. The average density of states as a function of energy [is] equal to a smooth monotonic function ... plus a sum of sinusoidal oscillations. The wavelength and amplitude of each oscillation are respectively correlated with the period and the stability of a periodic orbit of the system.
>
> *(Delos and Du 1988, p. 1450)*

In other words, there is a correspondence between the average quantum density of states and the periods and stabilities of the classical periodic orbits, which allows a calculation of the quantum quantities on the basis of these classical quantities. Furthermore, this correspondence holds whether the system is regular or chaotic.

In order to apply this semiclassical theory to the experimental spectroscopic measurements, which give the average oscillatory strength, not density of states,

[27] Again, further technical details and explanation can be found in both Delos and Du 1988 and Du and Delos 1988.

Delos and Du derived the following formula which relates the average oscillator-strength density to the finite-resolution Green's function:

$$\bar{D}f(E) = -\frac{2m_e(E_f - E_i)}{\pi\hbar^2}\mathrm{Im}\langle D\psi_i|\bar{G}_E|\psi_i\rangle, \tag{5.14}$$

where the left-hand side is the average oscillator strength, and $\bar{G}(E)$ is the finite resolution Green's function. With these equations they are now able to apply this new correspondence principle to explain the experimental "anomalous" spectral data. As they explain, these equations

> lead to a simple physical picture as well as a quantitative theory explaining how oscillations in the absorption spectrum are correlated with closed classical orbits. ... When the atom absorbs a photon, the electron goes into a near-zero-energy outgoing Coulomb wave. This wave propagates away from the nucleus to large distances. For $r \gtrsim 50a_0$ the outgoing wave fronts propagate according to semiclassical mechanics, and they are correlated with outgoing classical trajectories. Eventually the trajectories and wave fronts are turned back by the magnetic field; some of the orbits return to the nucleus, and the associated waves (now incoming) interfere with the outgoing waves to produce the observed oscillations.
>
> *(Du and Delos 1988, p. 1902)*

As the preceding discussion shows, closed orbit theory is a semiclassical theory that involves a thorough blending of classical and quantum ideas. To once again use the phrase popularized by Berry, it involves putting quantum flesh on classical bones, where here the classical bones are the closed orbits on which the quantum mechanical spectra are constructed.

In order to use closed orbit theory, one must first use classical mechanics to calculate the allowed orbits of a charged classical particle moving under the action of the combined Coulomb and magnetic field.[28] These closed orbits can exhibit a variety of loops and zig-zags before returning to the nucleus. It turns out that, of all the possible allowed closed orbits of an electron in such a field, only about sixty-five orbits are relevant to explaining the quantum spectrum (Du and Delos 1988, p. 1906). Which orbits are relevant, and how they explain the anomalous spectra as Du and Delos claim above, has been summarized in an intuitive way by Hans von Baeyer (a colleague of Delos's in the Physics Department at College of William and Mary):

> Delos's insight was to realize that interpreting the departing and arriving electron as a wave meant that its outgoing and incoming portions will inevitably display the symptoms of interference ... [T]he survival of some of these quantum mechanical waves and the canceling out of others result in only certain trajectories' being allowed for the electron in its classical cometlike ramblings far from the nucleus ... Once Delos established that only some

[28] This methodology should once again call to mind Dirac's reciprocal correspondence principle, according to which problems in quantum mechanics are used to guide the development of new results in classical mechanics in order that those classical results can then be used to solve the quantum problem.

trajectories are produced, he had effectively explained the new mechanism that caused the mysterious ripples [in the absorption spectrum above the ionization limit]. The Rydberg electron is allowed to continue to absorb energy, so long as that energy is precisely of an amount that will propel the electron to the next trajectory allowed by the interference pattern.

(von Baeyer 1995, p. 108)

It is worth emphasizing again that this explanation of the anomalous resonances in the spectra is not a purely quantum explanation, deducing the spectrum directly from the Schrödinger equation. Rather, it requires a careful blending of quantum and classical ideas: at the same time that the Rydberg electron is being thought of quantum mechanically as a wave exhibiting the phenomenon of interference, it is also being thought of fictionally as a particle following specific classical closed-orbit trajectories.

Despite the unorthodox hybridization of classical and quantum ideas in this explanation of the anomalous spectra, closed orbit theory has proven to be strikingly successful empirically. With these classical closed orbits, one can predict the wavelength, amplitude, and phase of these resonances to within a few percent, and furthermore, the predictions of this theory have proven to be in very close agreement with the data generated by numerous subsequent experiments on the absorption spectra of hydrogen, helium, and lithium atoms in strong magnetic fields.[29] This striking success of closed orbit theory shows that even in atomic physics, which is clearly under the purview of quantum mechanics, it is classical mechanics as developed through modern semiclassics that is proving to be the appropriate theoretical framework for tackling many of these quantum problems.

Closed orbit theory can not only explain the particular details of the experimental spectra, but can also explain why the earlier, lower-resolution data of Tomkins and Garton yielded a very orderly series of oscillations, while the later, higher-resolution data of Welge and colleagues revealed a wildly irregular series of oscillations. The explanation, once again, rests on a thorough mixing of classical and quantum ideas – specifically, a mixing of the quantum uncertainty principle with the fact that classical chaos is a long-time ($t \rightarrow \infty$) phenomenon that, on short time scales, can still look orderly. Because the low-resolution experiments involved only a rough determination of the energy, only the short-time classical dynamics is relevant to the spectrum. The high-resolution experiments, by contrast, involved a more precise determination of energy, and hence the longer time dynamics of the classical system is relevant. Since the long-time dynamics of a classical electron in a strong magnetic field is chaotic, this complexity manifests itself in the spectra.[30]

[29] See Granger (2001), Chapter 1 for a review.

[30] See Du and Delos (1988) for a more detailed explanation and for a picture of a typical chaotic trajectory of Rydberg electron in a combined Coulomb and diamagnetic field.

Ten years after closed orbit theory was introduced, Delos, Kleppner, and colleagues showed that, not only can classical mechanics be used to generate the quantum spectrum, but, even more surprisingly, the experimental quantum spectrum can be used to reconstruct the classical trajectories of the electron. Part of the reason this is surprising is that electrons do not, in fact, follow classical trajectories at all. Recognizing this tension, they write, "We present here the results of a new study in which semiclassical methods are used to reconstruct a trajectory from experimental spectroscopic data. When we speak of the 'classical trajectory of an electron,' we mean, of course, the path the electron would follow if it obeyed the laws of classical mechanics" (Haggerty *et al.* 1998, p. 1592). While the previous experiments could be used to establish the actions, stabilities, and periods of the closed orbits, they could not be used to determine the orbits themselves, that is, the electron positions as a function of time. In this paper, however they show how "by doing spectroscopy in an oscillating field, we gain new information that allows us to reconstruct a trajectory directly – without measuring the wave function and without relying on detailed knowledge of the static Hamiltonian" (Haggerty *et al.* 1998, p. 1592).

Their experiment involves examining the spectrum of a highly excited (that is, Rydberg) lithium atom in an electric field (a phenomenon known as the Stark effect). While the behavior of a hydrogen atom in an electric field is regular, the behavior of a lithium atom in an electric field can be chaotic. Using an extension of closed orbit theory, they were able to show that an oscillating electric field reduces the strength of the recurrences in the spectrum, that is, the heights of the peaks, in a manner that depends on the Fourier transform of the classical electron orbits in the static electric field. Hence, by experimentally measuring the Fourier transform of the motion for a range of frequencies, one can then take the inverse Fourier transform, and obtain information about the electron's orbits. Using this technique, they were able to successfully reconstruct from the experimental *quantum* spectra, two *classical* closed orbits of an electron in an electric field (these orbits are referred to as the "2/3" and "3/4" orbits, and are pictured in Figure 5.4).

They conclude, "Our experiment produces accurate, albeit low-resolution, pictures of classical trajectories important to the Stark spectrum of lithium" (Haggerty *et al.* 1998, p. 1595). Although they used the Stark spectrum of Rydberg lithium in this experiment, their method of extracting classical trajectories from quantum spectra can be applied to a variety of other systems. They emphasize that the limits they encountered in resolving these trajectories are experimental, not fundamental, being many orders of magnitude away from the limits imposed by Heisenberg's uncertainty principle. Hence, although these experiments are deriving pictures of classical trajectories from quantum spectra, these trajectories in no way undermine the uncertainty principle.

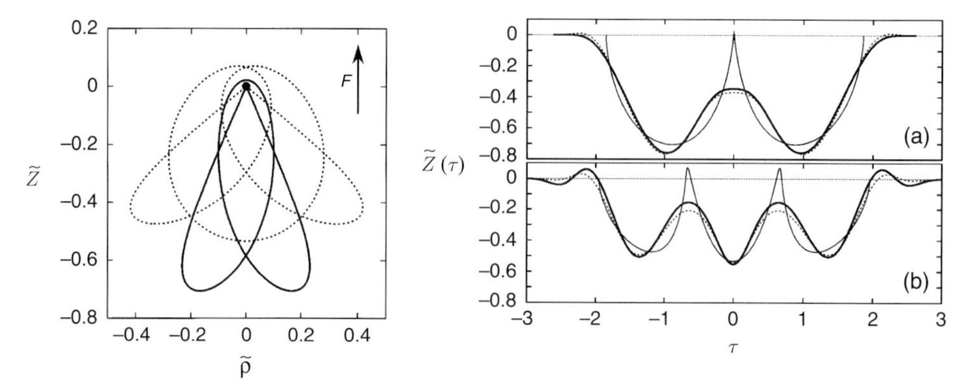

Figure 5.4 On the left: Two classical closed orbits of a Rydberg electron in an electric field; the solid line is the "2/3" orbit and the dotted line is the "3/4" orbit. The dark dot represents the nucleus of the atom. On the right: (a) the "2/3" orbit and (b) the "3/4" orbit. The light solid lines show the exact classical trajectories and the heavy lines are the experimentally reconstructed trajectories. Since the experimental frequency range is limited, the exact trajectories filtered through the experimental frequency window have also been included as the dashed lines for comparison. (From Haggerty *et al.* 1998, Figure 2 and Figure 4; courtesy of D. Kleppner).

Underlying these experiments is the following pressing but unspoken question: Given that classical mechanics is not true, and that electrons in atoms do not actually follow definite trajectories, how can one legitimately speak of experimentally measuring such trajectories from a quantum spectrum at all? In a more recent paper, Kleppner and Delos (2001) tackle this question of the reality of these trajectories head on. They write,

> Because of the power of the concept of periodic orbits, one might question as to what extent they "really exist" in the atom. Insight into this question was provided by a recent experiment at MIT, in which recurrence spectroscopy was used to measure the time dependence of an electron's motion along one of the closed orbits. To put it more precisely, recurrence spectroscopy was used to measure the time dependence of the fictitious classical trajectory that can be used as a calculational device to construct the quantum propagator. And to put it less precisely, recurrence spectroscopy showed how an electron in an atomic system would move in space and time if it obeyed classical physics.
>
> *(Kleppner and Delos 2001, p. 606)*

After discussing their experiment in more detail, however, Kleppner and Delos seem tempted by the view that these electron trajectories are more than mere fictions or calculational devices: They write, "These results lead us to question whether a trajectory should be described as truly 'fictitious' if one can measure its detailed properties" (Kleppner and Delos 2001, p. 610). The full realist claim, that electrons in atoms *really are* following these definite classical trajectories, would amount to a rejection of modern quantum mechanics and a violation of Heisenberg's uncertainty principle; and this is not something that semiclassical theorists, Kleppner and Delos included, intend to do.[31]

The more interesting philosophical question, in my view, is not whether these electron trajectories are real, but rather, how is it that certain fictions in physics can be so fruitful – indeed so much so that one can experimentally measure their detailed properties? What seems to be called for – given these experiments and the fertility of using classical trajectories in semiclassical mechanics more generally – is something less than a full-blown realism, yet more than a mere instrumentalism that dismisses them as nothing more than a calculational device. The task of articulating such an account shall be taken up in Chapter 6, after presenting one final case example of the fertility of using classical trajectories to explain quantum phenomena.

5.5 Wavefunction scars and periodic orbits

The fertility of using classical trajectories to investigate and explain quantum phenomena is not limited to the behavior of Rydberg atoms in strong electromagnetic fields. Classical structures such as periodic trajectories are also playing a central role in accounting for a quantum phenomenon known as wavefunction scarring. Wavefunction scarring was first discovered in a model system known as quantum billiards, which are the quantum analogs of classical billiard systems with chaotic geometries. As such, wavefunction scarring provides another example of the surprising new quantum phenomena that can occur when the corresponding classical system is chaotic, and even more surprisingly, how the quantum dynamics seems to retain the imprint of the underlying classical dynamical structures.

A quantum billiard system that has been at the forefront of semiclassical research on scarring is the so-called stadium billiard. This system consists of free-particle motion in a two-dimensional stadium-shaped enclosure. Classically, the typical trajectory in this system will bounce chaotically around the enclosure, passing through any region an infinite number of times with an infinite number of momentum directions (see Figure 5.5).

[31] There are, of course, consistent interpretations of quantum mechanics, such as Bohm's hidden variable theory, in which electrons do follow definite trajectories. Typically, however, these Bohmian trajectories are not the trajectories of classical mechanics. As an empirically equivalent formulation of quantum mechanics, Bohm's theory would nonetheless be able to account for any experimental results just as well as the standard interpretation. For those who are interested in Bohmian mechanics, a Bohmian approach to these diamagnetic Rydberg spectra has been carried out by Alexandre Matzkin, who concludes "*Individual* BB [de Broglie–Bohm] trajectories do not possess these periodicities and cannot account for the quantum recurrences. These recurrences can however be explained by BB theory by considering the *ensemble* of trajectories … although none of the trajectories of the ensemble are periodic, rendering unclear the dynamical origin of the classical periodicities" (Matzkin 2006, p. 1).

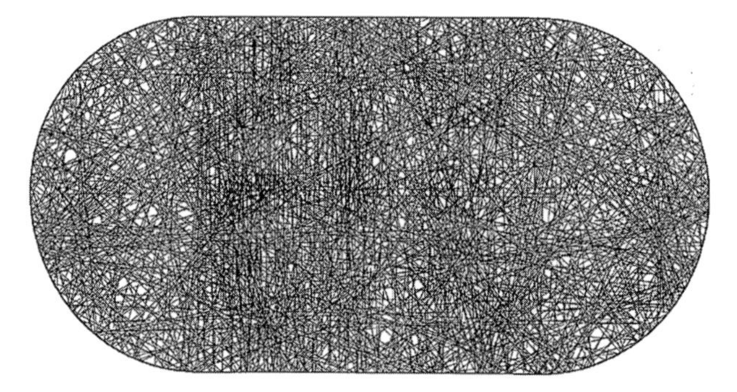

Figure 5.5 The stadium billiard system consisting of free-particle motion in a two-dimensional stadium-shaped enclosure. A typical chaotic trajectory is pictured here (courtesy of E. Heller).

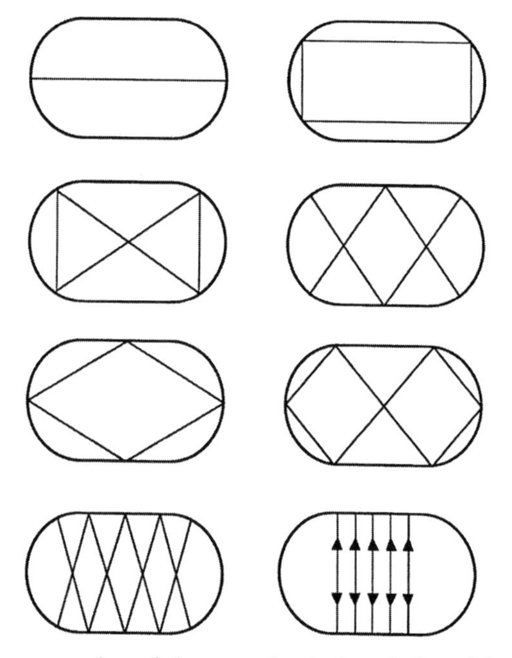

Figure 5.6 Some examples of the rare classical periodic orbits in the stadium billiard. All but the bottom right ("bouncing ball") orbits are isolated and unstable. (Reproduced from Figure 6 of Heller 1986 with kind permission of E. Heller and Springer Science and Business Media.)

There is, however, a small number[32] of trajectories that exactly repeat themselves, known as periodic orbits. Some examples of these rare short periodic orbits in the stadium billiard are the so-called "rectangular," "bow-tie," "V," and "double-diamond" orbits (see Figure 5.6).

[32] By "small number" I mean a set of measure zero (though there is an infinite number of them!).

When one considers the quantum-mechanical description of this same system, the expectation is that, locally, the wavefunctions should all look like a random superposition of plane waves (with fixed wavevector magnitude, but random amplitude, phase, and direction). This expectation was put in the form of a conjecture by Michael Berry (1983, p. 204; see also 1977), and is referred to as the "random eigenstate" hypothesis. Visually, such a random superposition of plane waves exhibits an irregular and diffuse pattern of probability density when plotted in coordinate space.[33] Surprisingly, however, numerical simulations of the quantum stadium billiard performed by Eric Heller in the early 1980s revealed that, rather than being irregular and diffuse, the probability density of many wavefunctions is strongly localized around the rare classical periodic orbits.[34] He dubbed this phenomenon "wavefunction scarring."

Wavefunction scarring is defined as the anomalous enhancement of quantum eigenstate intensity along unstable periodic orbits of a classically chaotic system (Kaplan 1999, p. R1).[35] If one plots the probability distribution of the wavefunction in coordinate space (a contour plot of $|\psi|^2$), one sees that there are certain eigenstates of the quantum stadium billiard whose wavefunction distributions are dominated by an individual short (i.e., few bounces) periodic orbit. Figure 5.7 shows three scarred eigenstates of the stadium billiard and the classical periodic orbits corresponding to these scars.

These images show quite vividly the way in which the quantum dynamics seems to retain the imprint of these classical dynamical structures. Despite the fact that these systems are in the semiclassical regime (i.e., high energy states), this phenomenon is not what one would expect on the basis of simple correspondence arguments: a classical probability distribution launched along one of these periodic orbits will show no such accumulation of density along the orbit. This has led one theorist to conclude that "the existence of these highly structured wavefunctions suggests that a reinterpretation of the correspondence principle is necessary, at least in this intermediate regime between the microscopic quantum and macroscopic classical worlds" (Jensen 1992, p. 314). As we saw in the last chapter, this is not Bohr's correspondence principle, but rather what was described in Chapter 1 as the *generalized* correspondence principle – the expectation that quantum and classical predictions should agree in the appropriate limit. Although scarring does not threaten this principle, it does show that the classical limit is far more subtle and complex than we might have first expected. The question I want to focus on here, however, is in what sense – if any – do these classical periodic orbits *explain* the wavefunction

[33] See O'Connor *et al.* (1987) for a discussion and pictures of superpositions of random plane waves.
[34] This phenomenon was subsequently observed experimentally in the lab by Sridhar (1991).
[35] An unstable orbit is one whose neighboring trajectories are diverging exponentially.

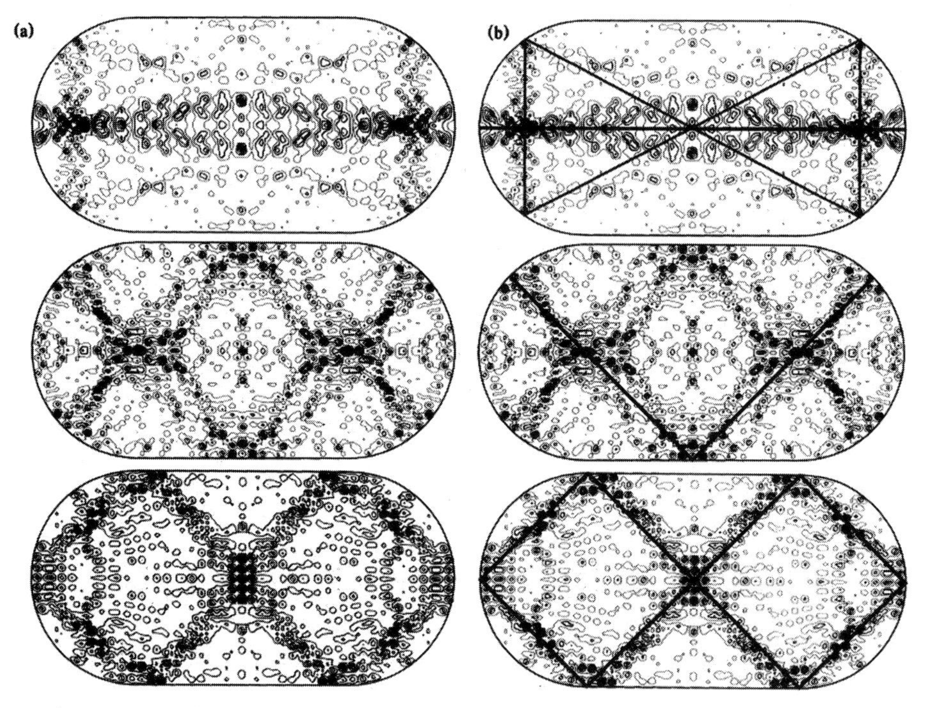

Figure 5.7 Left column (a) three eigenstates of the stadium billiard, and right column (b) the corresponding classical periodic orbit (solid line) associated with this scarred state superimposed. (Reproduced with permission from Heller 1984.)

scarring? This question becomes all the more pressing when one recalls that, according to quantum mechanics, there is no such thing as these classical trajectories at all.

In his original paper on wavefunction scars, Heller provided the following, now well accepted, theoretical explanation for their existence.[36] His explanation makes use of the "time-dependent picture" and exploits the fact that the classical evolution of Gaussian wavepackets can serve as arbitrarily accurate solutions of the time-dependent Schrödinger equation in the classical ($\hbar \rightarrow 0$) limit. He begins by asking us to consider the behavior of a Gaussian wavepacket, $\varphi(x,0)$, launched on one of these unstable classical periodic orbits (or more specifically a point in phase space associated with the periodic orbit), and watch the behavior of that wavepacket as it propagates forward in time $\varphi(x,t)$. Specifically, if we consider the autocorrelation of the wave packet, $\langle \varphi(t)|\varphi(0)\rangle$, that is the overlap of the wavepacket at some later time with the initial wavepacket, we see (in the left-hand side of Figure 5.8) that there is, first, a drop to zero as the wavepacket moves away from its initial position,

[36] The presentation here follows Heller (1984), Meredith (1992), and Kaplan (1999).

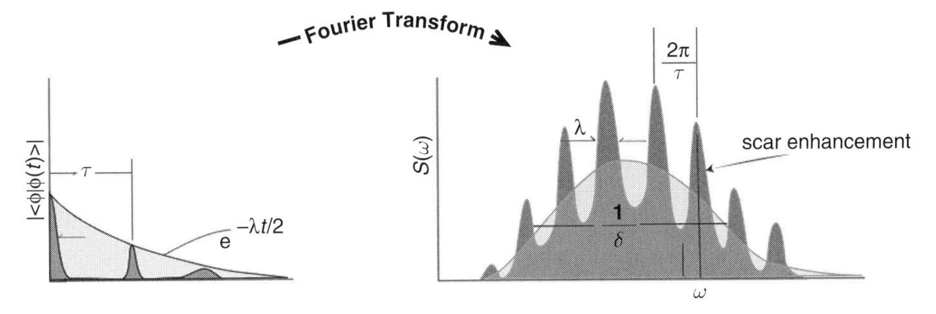

Figure 5.8 On the left, a plot of the autocorrelation of a wavepacket function of a wavepacket launched along an unstable periodic orbit as a function of time (where τ is the period of the particular periodic orbit). On the right, the Fourier transform of this autocorrelation function, where $S(\omega)$ is the local of states. (Adapted from Heller (personal communication).)

then second, a recurrence at the period of the classical orbit where it overlaps again, and then eventually a decrease in the strength of these recurrences with later multiples of the orbit's period due to the spreading of the Gaussian wavepacket as it falls off the periodic orbit. It is worth emphasizing that no such recurrences are predicted – or permitted – within the random eigenstate hypothesis. The rate of divergence is given by the classical Lyapunov (or stability) exponent λ of the periodic orbit, which measures its instability. One can then calculate the quantum spectrum by Fourier transforming this autocorrelation function. As we see in the right-hand side of Figure 5.8, the initial quick drop to zero gives rise to a wide energy envelope, and the recurrences at the period of the classical orbit give rise to oscillations in the envelope, with the width of those peaks given by λ.

The implications of these peaks for the probability distribution of the wavefunctions in phase space (and hence coordinate space) can be seen as follows. The wavefunction intensity in phase space is defined by the square of the absolute value of the overlaps of the wavefunction with Gaussian wavepackets $|\langle\varphi|\psi_n\rangle|^2$ centered at each point in phase space. If a particular point of phase space belongs to a periodic orbit then the overlap, and hence intensity, at that point will be larger than statistically expected. Furthermore, since the (absolute value squared of the) overlap of the wavefunction with the Gaussian at one time remains the same at all times, $|\langle\varphi|\psi_n\rangle|^2 = |\langle\varphi(t)|\psi_n\rangle|^2$, the wavefunction will have a larger intensity all along the periodic orbit; this is the scar on the wavefunction. The strength of the scar is (modulo some statistical fluctuations) determined by the purely classical Lyapunov exponent λ, which gives the width of the peak of the spectrum. If, on the other hand, the wavepacket is launched on a point that does not belong to a periodic orbit, then there will be no recurrences, no oscillations in the spectrum, and hence no localization of the wavefunction to that region.

On the one hand, there is no doubt that wavefunction scarring is a purely quantum phenomenon, insofar as it is essentially a phase interference phenomenon. Yet as we have just seen, the received explanation of scarring makes central reference to classical structures. In particular, there are two junctures at which classical structures are playing an explanatory role: first, classical periodic orbits are used to explain the behavior of the Gaussian wavepackets; that is, they are used to explain why wavepackets launched at certain points in phase space lead to recurrences in the autocorrelation function, while wavepackets launched at other points in phase space do not. Second, the classical Lyapunov exponent of the orbit is used to explain the strength of the scar; that is, why certain regions of phase space have an even greater accumulation of wavefunction intensity than others. If we accept that quantum mechanics is right, however, then there is no such thing as these periodic orbits or their stability exponents in quantum systems at all. The pressing philosophical question then becomes, how can these classical structures be explanatory if they do not exist? In an insightful article on the "culture of quantum chaos," M. Norton Wise and David Brock emphasize the importance of such an investigation. Regarding the phenomenon of wavefunction scarring they conclude,

> The attribution of cause and influence to classical orbits might seem to be only a way of talking about how the simulations function in assisting understanding, but they nevertheless provide the intuitive and theoretical explanation of the phenomena. Their status will require much more reflection.
>
> *(Wise and Brock 1998, p. 380)*

While I believe that these classical orbits are playing an essential explanatory role, I do not think that the sense in which they explain is causal. The challenge of providing this much-needed further analysis into the ontological status and explanatory power of classical structures in quantum phenomena shall be taken up in the next chapter.

5.6 Conclusion

Although it is typically believed that classical trajectories were banished from quantum systems with the downfall of the old quantum theory, recent work in semiclassical mechanics reveals that they still have a legitimate, though revised, role to play. In particular, as the research described in the previous sections shows, classical closed and periodic orbits have proven to be a fruitful framework for investigating, calculating, and even explaining quantum phenomena. The fertility of using semiclassical approaches lies in the fact that these methods can often provide physical insight into the dynamical structure of quantum systems, in a way that the fully quantum approaches do not. As researchers in this area have argued,

> [S]emiclassical methods are unrivaled in providing an intuitive and computationally tractable approach to the study of atomic, molecular, and nuclear dynamics. An important advantage of such methods is their ability to uncover in a single picture underlying structures that may be hard to extract from the profusion of data supplied by detailed quantum calculations.
>
> (*Uzer* et al. *1991, p. 42)*

Examples of this ability of semiclassical methods to uncover structural features of the quantum dynamics were seen both in the spectra of diamagnetic Rydberg atoms (Section 5.4), and in the intra-shell resonant states of the helium atom (discussed at the end of Section 5.3), which semiclassical analyses showed were not in fact localized along the so-called Wannier ridge, as had been believed.

Not only have classical trajectories proven to be a useful tool for investigating quantum systems, but even more remarkably, recent experiments in semiclassical mechanics have revealed new quantum phenomena that seem to retain the imprint of these classical dynamical structures. This sort of striking correspondence between, for example, particular features of the quantum spectrum and individual classical closed trajectories, suggests that there is an underlying continuity of dynamical structure between classical and quantum mechanics. As was seen in the case of wavefunction scarring, this continuity is often not what one would expect on the basis of naive correspondence arguments. In other words, it is not just that the quantum behavior is mimicking the classical behavior in the mesoscopic domain, but rather that particular classical structures are manifesting themselves in surprising ways in the quantum phenomena.

Many more examples of these correspondences between classical dynamical structures and quantum phenomena can be adduced.[37] For example, in condensed matter physics, mesoscopic conductors known as quantum dots can be constructed that are small enough to behave as a single (quantum) phase-coherent unit. It turns out that the experimentally measurable conductance properties of these quantum dots depend on the properties of the *classical* electron dynamics, more specifically on whether the classical dynamics is regular or chaotic[38]. Harold Baranger, Rodolfo Jalabert, and Douglas Stone (1993), who first discovered this effect, developed a new semiclassical theory analogous to Gutzwiller's formula to account for these results. Their semiclassical theory shows how the "quantum transmission coefficient determining the current through a quantum dot can ... be expressed through sums over pairs of phase-carrying trajectories" (Richter 2000, p. 9). That is, these hybrid semiclassical methods assume fictitiously that the electrons in the quantum

[37] For a review of some additional examples in mesoscopic physics see Richter (2000), Chapter 1.
[38] See Baranger *et al.* (1993) for further details.

dot are following classical trajectories, and then add quantum phases to these trajectories in order to recover the relevant quantum interference phenomena.[39]

Even more interesting for our point here, however, is the fact that subsequent experiments on the conductance properties of quantum dots revealed that not only do they depend on the classical dynamics in this universal way (differing for chaotic and nonchaotic geometries), but they also exhibit surprising non-universal features that depend on the individual classical periodic orbits belonging to the particular quantum dot.[40] More specifically, these experiments revealed that there are strong correlations in the heights of the conductance peaks, whose explanation depends on the particular classical periodic orbits in the quantum dot. Steven Tomsovic and colleagues, who first provided an account of these correlations explain the phenomenon as follows:

> [T]he particular correlations in a given dot are not universal but rather involve detailed information about the dot ... [W]e use semiclassical techniques to derive a relation between the quantum conductance peak height and the classical periodic orbits in the dot. The main result is that as a system parameter varies – the magnetic field, for instance, or number of electrons in the dot ... – the interference around each periodic orbit oscillates between being destructive and constructive. When the interference is constructive for those periodic orbits which come close to the leads used to contact the dot, the wavefunction is enhanced near the leads, the dot-lead coupling is stronger, and so the conductance is larger.
>
> *(Narimanov* et al. *2001, pp. 1–2)*

In other words, the heights of the conductance peaks are explained by appealing to interference properties along the *fictitious classical* periodic orbits in the quantum dot. Once again, although these classical trajectories are playing an explanatory role, they are not interpreted realistically as the paths the electrons are actually following.

This example, like the other case studies discussed in much more detail in the preceding sections, underscores the following three lessons: First, there is a variety of *quantum* phenomena ranging from atomic physics to condensed matter physics for which semiclassical mechanics – not straight quantum mechanics – provides the appropriate theoretical framework for investigating, calculating, and explaining these phenomena. Second, these semiclassical methods and explanations involve a thorough hybridization of classical and quantum ideas (such as constructive and destructive quantum interference along classical periodic orbits). Far from being incommensurable theoretical concepts, these classical and quantum notions can be combined in both empirically adequate and conceptually fruitful ways. And finally, these classical structures (such as periodic orbits), are not simply useful calculational devices, but are actually manifesting themselves in surprising ways in the quantum experiments, as we saw in both the case of wavefunction scarring and the

[39] A detailed review of these semiclassical methods in mesoscopic condensed matter systems can be found in Jalabert (2000) and references therein.

[40] See Narimanov *et al.* (1999) and (2001) for details, and references therein.

case of the unexpected oscillations in the diamagnetic Rydberg spectrum. This speaks to a much richer continuity of dynamical structure across classical and quantum mechanics than is usually recognized.

A striking feature of all the examples discussed in this chapter is that they involve using classical mechanics to investigate, calculate, and come to a deeper understanding of quantum phenomena. This is surprising in so far as quantum mechanics – not classical mechanics – is the more fundamental physical theory. The usual way of expressing the classical limit suggests that the task of understanding the relationship between these two theories involves showing how quantum mechanics can be used to investigate, calculate, and come to a deeper understanding of classical phenomena. In other words, the task is supposed to be that of showing how quantum mechanics can reproduce the classical phenomena, not the other way around. This dilemma is an example of what Daniel Kleppner has called the "bow-stern enigma." He writes,

> The term has its origin in a canal voyage I made with some friends. Upon arriving at our boat we were confronted with an embarrassing but apparently fundamental problem: deciding which end of the boat was the bow and which the stern. The problem, however, turned out to be merely superficial. The countryside was so delightful that it made little difference which way we traveled: The scenery was interesting in either direction.
>
> *(Kleppner 1991, p. 11)*

The suggestion, then, is that whether one is using quantum mechanics to investigate classical quantities or classical mechanics to investigate quantum quantities, there are important and interesting things to learn about the relationship between these two theories in either direction.

Although the many new scientific insights and discoveries that semiclassical mechanics brings are, of course, important in their own right, our interest here is specifically in what implications this research has for debates in the philosophy of science. In what follows, I want to focus on two challenges that I see semiclassical research posing for the philosophy of science. The first is whether there is a philosophical account of scientific explanation that can make sense of these appeals to classical structures in explaining quantum phenomena. I argue that philosophers of science should take the actual explanatory practices of scientists seriously, and rather than rejecting a purported explanation as no explanation at all (simply because it does not fit our preconceived philosophical views about explanation), we should instead see if a new philosophical account is called for. I shall argue that there is indeed a new mode of explanation in the examples described here, and the task of developing a philosophical account of explanation appropriate to these examples is taken up in the next chapter.

The second challenge that this semiclassical research raises is to the adequacy of our current philosophical frameworks for thinking about intertheory relations.

More specifically, this research suggests a more subtle and intricate relation between classical and quantum mechanics than the traditional frameworks of reductionism and theoretical pluralism allow. Articulating the nature of this challenge, and how we, as philosophers of science, are to respond to it, is the subject of the final chapter.

6

Can classical structures explain quantum phenomena?

> … there is something in this more than natural, if philosophy could find it out.
>
> *Shakespeare,* Hamlet, *Act 2 Scene 2*

6.1 Introduction

The philosophical importance of semiclassical appeals to classical structures in explaining quantum phenomena was first recognized by Robert Batterman (1993; 1995; 2002). More generally, Batterman has argued that when it comes to explaining phenomena in the asymptotic domain between two theories, the fundamental theory describing that domain is often explanatorily deficient, and an adequate explanation must make essential reference to the less fundamental theory. For example, in discussing the relation between classical and quantum mechanics, Batterman writes, "There are many aspects of the semiclassical limit of quantum mechanics that cannot be explained purely in quantum mechanical terms, though they are in some sense quantum mechanical … [T]hese quantum mechanical features require reference to classical properties for their full explanation" (Batterman 2002, pp. 109–10). He has extended these arguments to other theory pairs with singular limits as well, arguing that in optics, for example, one finds explanations of wave-theoretic phenomena that make an essential and ineliminable appeal to ray-theoretic structures such as caustics.[1]

More recently, Batterman's arguments have been the subject of considerable criticism from figures such as Clifford Hooker (2004), Michael Redhead (2004), and Gordon Belot (2005). These criticisms have tended to cluster around the following two assumptions: First, in saying that a classical structure (such as a

[1] In geometrical optics, a caustic is the envelope of a family of light rays, that is, a bright boundary produced by the rays coming together, on the other side of which the light intensity goes to zero.

trajectory or caustic) *explains*, one is thereby committed to the claim that the structure *exists*.[2] And, second, in saying that quantum mechanics is explanatorily deficient and that a full explanation requires reference to classical structures, one is thereby committed to the claim that there does not exist a purely quantum-mechanical explanation for the phenomenon in question.[3] Indeed these two assumptions have essentially been used as modus tollens arguments against Batterman. Despite the prima facie plausibility of these assumptions, however, I want to argue that they are in fact mistaken: one can take a structure to explain without taking that structure to exist, and one can maintain that, even though there may be a purely quantum-mechanical explanation for a phenomenon, that explanation – without reference to classical structures – is in some sense deficient.

Although my views are largely sympathetic with Batterman's, the kinds of semiclassical explanations we are interested in, and hence the accounts of scientific explanation we each develop, are quite different. Batterman is specifically interested in what are called *universal* phenomena, that is, phenomena which reappear in a variety of distinct sorts of physical systems, and for which the particular details of those systems are largely irrelevant for the explanation.[4] To clarify how explanations of universal phenomena are different, Batterman (1992; 2002) makes the following useful distinction between two kinds of why-questions: "A type (i) why-question asks for an explanation of why a given instance of a pattern obtained. A type (ii) why-question asks why, in general, patterns of a given type can be expected to obtain" (Batterman 2002, p. 23). His account of "asymptotic explanations" is specifically concerned with answering type (ii) questions. While I think Batterman is quite right to call attention to the importance of universal phenomena, which have been largely overlooked in the philosophy of science literature, I think there are important type (i) questions to ask about semiclassical mechanics as well. So in the case of scarring I am interested in why some particular wavefunction morphology occurs (e.g., why the bow-tie scar); or, in the case of the absorption spectrum of diamagnetic Rydberg atoms, why some particular sequence of sharply defined peaks occurs. These are phenomena that depend on the particular details of the physical system in question: change the shape of the billiard and a different set of scars will appear; for Rydberg atoms, change the value of the magnetic or electric field and there will be a different sequence of peaks. The importance of explaining these non-universal phenomena was also seen in the example of quantum dots,

[2] For examples of this first sort of objection see Redhead (2004, p. 530) and Hooker (2004, pp. 446–8).

[3] This assumption underlies Belot (2005); his strategy is to show that "one is able to prove the results required to explain the phenomena in question without recourse to less fundamental theories" (p. 152), hence reference to the less fundamental (here classical) structures is eliminable.

[4] See Batterman's (2002) Section 2.2 for a helpful introduction to universality.

discussed briefly in Section 5.6, in which the correlations in the heights of the conductance peaks was shown to depend in a non-universal way on detailed information about the quantum dot.

In what follows, I am concerned specifically with providing a philosophical account of these type (i) explanations that were encountered in the last chapter, that is, explanations of particular quantum phenomena that appeal to particular classical structures. I argue that these explanations do not require us to reify these classical structures – at least not in a naive way. More generally, I shall defend the view that fictional classical structures can be genuinely explanatory by developing a new model-based account of scientific explanation. Finally, I shall argue that these semiclassical explanations are indispensable in the sense that they can often provide a deeper understanding of the physical phenomena than the purely quantum-mechanical explanations do.

6.2 The reality and explanatory power of classical trajectories

The case studies discussed in the previous chapter illustrate the surprising ways in which classical structures are manifesting themselves in quantum phenomena. The fertility of using classical trajectories (especially closed orbits and periodic orbits) to investigate, calculate, and explain quantum phenomena raises the following two questions: First, what is the ontological status of these classical trajectories? And second, is this appeal to classical trajectories in explaining quantum phenomena legitimate?

In one sense, the question of the ontological status of these classical trajectories has the following straightforward answer: classical trajectories do not exist in quantum systems; they are fictions.[5] In the case of atoms in strong external fields, the Rydberg electron simply is not traveling in a localized manner along these closed orbits, and in the case of scarring, there is no localized particle reflecting off the boundary of the billiard along the path of the periodic orbits. A classical trajectory requires a simultaneously well-defined position and momentum, which on the standard interpretation is in conflict with Heisenberg's uncertainty principle. Well-confirmed

[5] In another sense, the issue of the ontological status of these classical trajectories is perhaps less straightforward. I have not, for example, discussed the possibility that these classical trajectories might exist as abstract structures without concrete realizations. Such a position might be attractive for those who want to maintain a realist view about explanation (i.e., only real things can explain). There is, in fact, an important parallel literature in the philosophy of mathematics that seeks to use indispensability in scientific explanation as an argument for platonism in mathematics. I am grateful to Alan Baker (personal communication) for bringing this literature to my attention, and a nice entrée into this literature on the explanatory power of mathematics can be found in Baker (2005). Regrettably an adequate discussion of these issues and connections will have to await a future work. I will say this much, however: I am reluctant to view these trajectories as purely mathematical objects since they carry a lot of *physical* information about both the classical and quantum dynamics. Hence I believe they are best understood as *dynamical* structures, and in the case of periodic orbits in quantum systems, they are dynamical structures without concrete realizations.

quantum experiments, such as the infamous two-slit experiment, also highlight the obstacles one faces in attributing classical trajectories to quantum particles.[6]

Nonetheless, one can begin to understand why quantum physicists such as Delos and Kleppner take these classical trajectories to be more than just calculational devices. In particular, there are two features of the semiclassical research described in the previous chapter that suggest that these classical trajectories should be thought of as more than "*mere* fictions." First, one can measure the detailed properties of these trajectories directly from the quantum data, and second, these closed and periodic orbits seem to be playing a fundamental explanatory role in accounting for quantum phenomena. Recall that, as Main and collaborators write, in the case of diamagnetic Rydberg atoms, "the regular type resonances can be physically rationalized and explained by *classical* periodic orbits of the electron on closed trajectories starting at and returning to the proton as origin" (Main *et al.* 1986, pp. 2789–90). Similarly in the case of quantum billiards, periodic orbits and their stability exponents are used to explain both why certain regions of phase space have a higher than expected accumulation of wavefunction density and their particular strength.

In discussing the ontological status of classical trajectories in quantum phenomena, it is helpful to distinguish between a calculational device and a model. Although they are both "fictions," a model goes beyond a calculational device insofar as it purports to capture adequately certain features of the empirical phenomenon. This is not to say that models cannot also function as calculational devices. Recognizing that classical trajectories can function as models of certain features of the quantum dynamics allows us to make philosophical sense of their rich fertility and explanatory power in semiclassical research – indeed it can make sense of this fertility in a way that dismissing them as a mere calculational devices cannot.

A resolution of this tension, concerning the ontological status of these closed and periodic orbits, requires one to take a broader perspective of the relationship between classical and quantum mechanics. Not only was quantum mechanics built on the framework of classical mechanics (as we saw in Chapters 3 and 4), but in the semiclassical regime in which Rydberg atoms and scarring occur, one would expect classical mechanics, which describes quite accurately our everyday macroscopic world, to be able to model certain features of the quantum dynamics. As I emphasized earlier, however, classical structures can model quantum dynamics in far more subtle and complex ways than merely "approximating" (i.e., in ways other than having a quantum particle approximately following a classical path in a localized way). This more complex modeling relation can be seen, for example, in the case of

[6] There are, of course, (nonlocal and contextual) interpretations of quantum mechanics, such as Bohmian mechanics, in which particles do follow definite trajectories; however, these trajectories will not typically be the trajectories of classical mechanics.

Gutzwiller's trace formula (discussed in Section 5.2), which uses a sum of classical periodic orbits to model the quantum density of states. The fertility of using classical concepts to model quantum phenomena lies in the fact that there is a continuity of dynamical structure between classical and quantum mechanics, and it is this dynamical structure, common to both theories, which is manifesting itself in these semiclassical experiments. Although there is no particle following these classical closed and periodic orbits, these trajectories nonetheless legitimately model certain features of the quantum dynamics in the semiclassical regime.

More rigorously, the justification for using classical trajectories as models of certain aspects of the quantum dynamics lies in Delos's closed orbit theory (discussed in Section 5.4) and Gutzwiller's periodic orbit theory (discussed in Section 5.2). These theories, which are grounded in Feynman's path integral formulation of quantum mechanics, spell out in detail the way in which these classical dynamical structures are to be properly connected up with quantum quantities.

Once one accepts that these trajectories are fictions, more specifically fictions in the sense of models, then the next philosophical task becomes making sense of how these classical structures can carry any explanatory force (either in explaining the oscillations in the quantum spectrum above the ionization limit, or in explaining the enhanced probability density of a wavefunction in certain regions of a quantum billiard) if they do not, strictly speaking, exist.

The intuition that a structure must exist in order to explain comes from the two most widely received accounts of scientific explanation: Carl Hempel's deductive–nomological (D–N) explanation and Wesley Salmon's causal explanation. Recall that according to the D–N account, a scientific explanation involves deducing the phenomenon to be explained (the explanandum) from the relevant laws and initial conditions (the explanans). A requirement of the D–N account is that the sentences constituting the explanans must be true (Hempel 1965, p. 248); hence an explanation that makes reference to fictional entities (typically) fails to meet this requirement.

Similarly, on the causal account of explanation, whereby to explain a phenomenon or event is to identify the causal mechanisms leading to that event (e.g., Salmon 1984), fictional structures also cannot explain. Obviously A cannot be the cause of B if A does not exist.[7] Although causal language is sometimes used in popular accounts, such as "periodic orbits that are not too unstable cause scarring" (Heller and Tomsovic 1993, p. 42), I think that it is clear that the sense in which physicists take these classical trajectories to explain certain aspects of the quantum phenomena is not causal.

The recognition that these classical structures are functioning as models, rather than mere calculational devices, provides a framework for making sense of their

[7] Causation by omission notwithstanding.

explanatory power. In particular, these closed and periodic classical orbits can be said to explain features of the spectral resonances and scarring insofar as they provide a semiclassical model of these phenomena. I shall refer to this type of scientific explanation as a *model explanation*.[8]

Models are used in a variety of ways in scientific practice; they can, for example, function as proto-theories, pedagogical devices, or as tools for generating and testing hypotheses.[9] In what follows I defend the view that – in some cases – models can perform an *explanatory* function as well.[10] Despite the widespread use of models to explain phenomena in physics, as well as the vast philosophical literature on models and explanation individually, remarkably little has been said about the explanatory function of models. I shall begin by briefly reviewing three different proposals for how models can genuinely explain. These are Ernan McMullin's (1978; 1985) "causal model explanations," Mehmet Elgin and Elliott Sober's (2002) "covering-law model explanations," and Carl Craver's (2006) "mechanistic model explanations." Abstracting from these three proposals, I will give a general account of what it is that makes an explanation a *model explanation*, and articulate the conditions under which it is reasonable to take such models as being genuinely explanatory. Finally, I shall show that none of these three types of model explanations can capture the sense in which classical structures are explaining the quantum phenomena in semiclassical research. Drawing on the work of R. I. G. Hughes (1989; 1993), I propose a new type of model explanation that I call "structural model explanation," which I think does adequately capture the way in which classical trajectories can be said to explain quantum phenomena.

6.3 Three kinds of model explanations

One of the earliest defenses of the view that models can be genuinely explanatory comes from McMullin's work on what he calls "hypothetico-structural" (or HS) explanations and the status of idealizations in science. In a hypothetico-structural explanation, one explains the properties of a complex entity by postulating an underlying structural model, whose features are causally responsible for those properties to be explained. He writes, "The structure underlying such an explanation

[8] Let me emphasize from the outset that model explanations are not intended to replace D–N or causal explanations; rather, they are meant to complement them. Scientists in their practice make use of a wide range of types of explanation, and in my view it is a mistake to try to reduce all scientific explanation to a unitary account.

[9] William Wimsatt (1987) provides a nice taxonomy of the various ways in which scientific models can be false and the various functions that such false models can serve in scientific research. Regrettably he does not talk about the explanatory function of false models, which is my focus here.

[10] Obviously not all models are explanatory. For example, one would not say that Descartes' vortex model or Ptolemy's epicyclic model actually explain the motion of the planets. It would be a mistake, however, to conclude from this that *no* models are explanatory. The conditions under which we can consider a model to be genuinely explanatory will be discussed below.

is often called a physical (or a theoretical) 'model,' since the explanation is a hypothetical one" (McMullin 1978, p. 139). He notes that HS explanations are usually metaphorical and tentative, and that "a good structural model will display resources for imaginative extension over a considerable period" (McMullin 1978, pp. 145–6).

McMullin explicitly distinguishes this type of explanation from nomothetic explanations, and although HS explanations may involve laws of nature, "the explanatory character of the model comes not just from the laws governing the constituents of the model but also from the structure in which these entities are combined" (McMullin 1978, pp. 146–7). For McMullin, these model explanations are ultimately to be understood as a species of causal explanations: "Such explanations are causal, since the structure invoked to explain can also be called the cause of the feature being explained" (McMullin 1978, p. 139). For this reason I think McMullin's account is best called a *causal model explanation*.[11]

In this 1978 paper, McMullin does not address the question of how it is that a hypothetical and metaphorical model can legitimately be said to explain at all. For his answer to this question we must look to another paper of his on Galilean idealization, where he writes, "Our interest here lies in the use of models as idealizations of complex real-world situations, the 'falsity' this may introduce into the analysis, and the ways in which this 'falsity' may be allowed for and even taken advantage of" (McMullin 1985, p. 258). He continues,

> Every theoretical model idealizes, simplifies to some extent, the actual structure of the explanandum object(s). It leaves out of [the] account features deemed not to be relevant to the explanatory task at hand. Complicated features of the real object(s) are deliberately simplified in order to make theoretical laws easier to infer, in order to get the process of explanation under way.
>
> *(McMullin 1985, p. 258)*

McMullin then goes on to provide a taxonomy of the various ways in which scientific models can involve idealizations:

> Idealization enters into the construction of these models in two significantly different ways. Features that are known (or suspected) to be relevant to the kind of explanation being offered may be simplified or omitted in order to obtain a result ... On the other hand, the model may leave features unspecified that are deemed irrelevant to the inquiry at hand.
>
> *(McMullin 1985, p. 258)*

He argues that it is important to distinguish these two types of idealization, for they both play an important role in the subsequent process of legitimating the models and showing that they are genuinely explanatory. For McMullin, the justification of

[11] Although this label regrettably does not capture McMullin's emphasis on structures, I would nonetheless like to reserve the term "structural explanation" for a distinctive type of explanation that is neither nomothetic nor causal, as will be discussed below.

the model proceeds through a process that he calls *de-idealization*. That is, one goes through a process of "adding back in" those features that were omitted, or de-simplifying those assumptions that were included, but only included in an over-simplified way.[12] It is through this (sometimes lengthy) process of de-idealizing, that McMullin sees metaphoric models as laying out a research program. He notes that for such a process to work, the original model must successfully capture the real structure of the object of interest. When this process of de-idealizing can be given a theoretical justification, then McMullin concludes that the model can be counted as genuinely explanatory. When, on the other hand,

> techniques for which no theoretical justification can be given have to be utilized to correct a formal idealization, this is taken to count against the explanatory propriety of that idealization. The model itself in such a case is suspect, no matter how good the predictive results it may produce.
>
> *(McMullin 1985, p. 261)*

McMullin is well known for taking this account of model explanation one step further, and arguing that when, in this process of de-idealizing a model, new discoveries are made and even more experimental data can be accounted for, then we have strong (though not conclusive) evidence for the existence of the structures postulated by the model (McMullin 1985, p. 262; and especially McMullin 1984).

For my purposes here, however, I want to bracket this issue of realism, and instead simply draw attention to McMullin's following three claims: first, idealized models can be genuinely explanatory; second, when they do explain, what they offer is a species of causal explanation (namely the structures postulated in the model are taken to *cause* the relevant observed features of system to be explained); and third, we are justified in taking the model as genuinely explanatory when a theoretical justification can be given for each step in the de-idealization process, that is, the de-idealization does not simply involve an ad hoc fitting of the model to empirical data.

More recently, Elgin and Sober (2002) have picked up on McMullin's theme of idealization, and have also defended the view that scientific models can explain. The example that they discuss is a class of models in evolutionary biology known as "optimality models"; these models describe the value of a trait that maximizes fitness, given a certain set of constraints. They argue that despite the idealizations involved in these optimality models, they are nonetheless genuinely explanatory. They write,

> Optimality models contain idealizations; they describe the evolutionary trajectories of populations that are infinitely large in which reproduction is asexual with offspring always

[12] Later in this same article, McMullin offers a nice illustration of this process using Bohr's model of the atom.

resembling their parents, etc. ... We want to argue that optimality models are explanatory despite the fact that they contain idealizations.

(Elgin and Sober 2002, p. 447)

The sort of model explanations that Elgin and Sober consider are a species of covering-law causal explanations. In order to qualify as a *covering-law* causal explanation, first, the explanans must describe the cause (or causes) of the explanandum; second, the explanans must cite a law of nature; and third, all of the explanans propositions must be true. The further condition usually added to a covering-law explanation, namely that the explanans explains the explanandum by entailing it or conferring probability on it, is not satisfied by covering-law model explanations, they argue, because it does not make sense to talk about the probability of an idealized circumstance (the one described in the model) that does not obtain in the real world.[13]

What makes Elgin and Sober's account of model explanation different from McMullin's is that, for them, the explanans must both cite a law of nature and be entirely true. For McMullin's causal model explanations, by contrast, the explanans need not be literally true (the structures appealed to could be metaphorical), and the explanans need not cite a law of nature. Hence to distinguish Elgin and Sober's account of model explanations, I shall refer to it as "*covering-law model explanations*."[14]

Elgin and Sober also recognize the need for distinguishing those model explanations that can be understood as being genuinely explanatory from those that are not. In a manner similar to McMullin's, their account proceeds by way of a de-idealization analysis. They write, "The idealizations in a causal model are *harmless* if correcting them wouldn't make much difference in the predicted value of the effect variable. Harmless idealizations can be explanatory" (Elgin and Sober 2002, p. 447). That is, on their view, a model can be explanatory only so long as the idealizations in the model are harmless.

Yet a third account of model explanation has been given by Craver in the context of models in the neurosciences. According to Craver, "models are explanatory when they describe mechanisms. Perhaps not all explanations are mechanistic. In many cases, however, the distinction between explanatory and non-explanatory models is that the latter, and not the former, describe mechanisms" (Craver 2006, p. 367). Mechanistic explanations are a kind of constitutive explanation, in which the

[13] Sober (personal communication) has subsequently clarified that the difficulty is specifically with conditionalizing on an idealization that involves an *impossible* situation (such as an infinite population size). Even if one grants that the probability of an event conditionalized on an impossible state of affairs has a well-defined value, it still poses a problem for the inferential link between explanans and explanandum.

[14] Note that "covering-law model explanation" should be distinguished from "covering-law model *of* explanation"; while in the latter "model" just means "an account," in the former "model" is meant to convey the fact that the explanans makes essential reference to a scientific model.

behavior of a whole is explained in terms of the operation and interaction of the mechanism's parts. His version of model explanations is then best described as *mechanistic model explanations*.

Craver places rather stringent conditions on when a mechanistic model is to count as genuinely explanatory. He writes, "In order to explain a phenomenon, it is insufficient merely to characterize the phenomenon and to describe the behavior of some underlying mechanism. It is required in addition that the components described in the model should correspond to components in the mechanism in [the target system] T" (Craver 2006, p. 361). Craver elaborates on this requirement by drawing a distinction between "how-possibly models" and "how-actually models." A how-possibly model, he explains, merely describes "how a set of parts and activities might be organized such that they produce the explanandum phenomenon" (Craver 2006, p. 361). How-actually models, by contrast, describe the "real components, activities, and organizational features of the mechanism that in fact produce the phenomenon" (Craver 2006, p. 361). In other words, for a mechanistic model to count as genuinely explanatory, it must correctly reproduce the *actual* mechanisms.

Craver's requirement that "a mechanistic explanation must begin with an accurate and complete characterization of the phenomenon to be explained" (Craver 2006, p. 368) drastically limits those models that can be counted as explanatory, since very few scientific models provide a complete and accurate characterization of the phenomenon to be explained.[15] Nonetheless, Craver argues that there are some models in the neurosciences that do meet these requirements, and when they do, they are genuinely explanatory models.

6.4 A general account of model explanations

The wide variety of different kinds of models employed in the sciences suggests that there may also be a wide variety of different kinds of model explanations. Indeed so far we have encountered three different kinds of model explanations: McMullin's causal model explanations in the natural sciences (excluding, for the most part, mechanics),[16] Elgin and Sober's covering-law model explanations in the context of evolutionary biology, and Craver's mechanistic model explanations in the neurosciences. Thus even after one has identified an explanation as a model explanation, there still remains the question of what *kind* of model explanation it is.

Despite the differences among these various characterizations of model explanations, I believe that there is a set of core features that unites them all as model

[15] Indeed Craver's requirements sound more like a theoretical description than a model.

[16] In many places in McMullin's writings he notes that the fields of mechanics and microphysics are special, and that these forms of explanation and inference do not hold in these fields as clearly as they do in other areas of science such as geology, astrophysics, chemistry, and cytology (see, for example, McMullin 1978, p. 147).

explanations. First, what makes them all examples of *model* explanations is that the explanans in question make essential reference to a scientific model, and that scientific model (as is the case with all models) involves a certain degree of idealization and/or fictionalization. Second, a general characterization of model explanations requires giving an account of how these models can be genuinely explanatory. My answer here draws on a suggestion made by Margaret Morrison, who writes, "The reason models are explanatory is that in representing these systems, they exhibit certain kinds of structural dependencies" (Morrison 1999, p. 63). Unfortunately, however, Morrison does not indicate how this suggestion is to be fleshed out into a philosophical account of scientific explanation, nor how it will be able to distinguish genuinely explanatory models from those phenomenological models that merely "save the phenomena."

There is, however, a similar account of scientific explanation that has been worked out in much greater detail by James Woodward. According to Woodward, an explanation can be understood as providing "information about a pattern of counterfactual dependence between explanans and explanandum" (Woodward 2003, p. 11). He fleshes out this idea of counterfactual dependence in terms of what he calls "what-if-things-had-been-different questions," or "w-questions" for short. That is, "the explanation must enable us to see what sort of difference it would have made for the explanandum if the factors cited in the explanans had been different in various possible ways" (Woodward 2003, p. 11; see also Woodward and Hitchcock 2003).

While I think that Woodward's account is largely right, where I wish to part company with his view is in his construal of this counterfactual dependence along strictly manipulationist or interventionist lines. Woodward takes one of the key features of this pattern of dependence to be that it, in principle, permits one to *intervene* in the system in various ways. It is precisely this manipulationist construal that restricts Woodward's account of scientific explanation to specifically *causal* explanations.[17] As I will argue in more detail below, I think it is a mistake to construe all scientific explanation as a species of causal explanation, and more to the point here, it is certainly not the case that all model explanations should be understood as causal explanations. Thus, while I shall adopt Woodward's account of explanation as the exhibiting of a pattern of counterfactual dependence, I will not construe this dependence narrowly in terms of the possible causal manipulations of the system.

[17] Woodward in fact admits that perhaps not all scientific explanations are causal explanations, and allows that his theory may be extended in the way I am suggesting. I will return to Woodward's discussion of this possibility in Section 6.5 below.

So far I have articulated two key features of a general account of model explanations: first, the explanans must make essential reference to a scientific model, and second, that model must explain the explanandum by showing how there is a pattern of counterfactual dependence of the relevant features of the target system on the structures represented in the model. That is, the elements of the model can, in a very loose sense, be said to "reproduce" the relevant features of the explanandum. Furthermore, as the counterfactual condition implies, the model should also be able to give information about how the target system would behave, if the structures represented in the model were changed in various ways.

In addition to these two requirements, a third condition that an adequate model explanation must satisfy is that there must be what I call a further "justificatory step." Very broadly, we can understand this justificatory step as specifying what the domain of applicability of the model is, and showing that the phenomenon in the real world to be explained falls within that domain. This justificatory step is intended to call explicit attention to the detailed empirical or theoretical process of demonstrating the domain of applicability of the model. In other words, it involves showing that it is a good model, able to adequately capture the relevant features of the world (where "relevant" is determined by which questions the model is specifically trying to answer).[18]

Although the details of this justificatory step will depend on the details of the particular model explanation in question, there are typically two general ways in which such a justification can proceed. In the first case, the justification might proceed "top down" from theory; that is, one might have an overarching theory that specifies what the domain of applicability of the model is – where and to what extent the model can be trusted as an adequate representation of the world.[19]

More frequently, however, there is no such overarching justificatory theory, and the justification must instead proceed from the ground up, through various empirical investigations. We can see the importance of this latter sort of justificatory step in the three examples of model explanations already discussed.[20] For example, Elgin and Sober's analysis that the idealizations employed in the evolutionary model are "harmless" seems to be an example of this sort of bottom-up justificatory procedure. Such a bottom-up de-idealization procedure of the sort that Elgin, Sober, and

[18] This justificatory step does not simply involve showing that the model is empirically adequate. As I will discuss in more detail below, it is to be understood as playing a role analogous to Hempel's condition of truth (which he refers to as the "empirical condition of adequacy"), insofar as the justificatory step is intended to rule out as explanatory those models that we know to be merely phenomenological.

[19] I will argue below that the justificatory step in the case of classical trajectories explaining quantum phenomena falls into this first category, where the overarching theory is given by Gutzwiller's periodic orbit theory or Delos's closed orbit theory.

[20] Even more typically (and perhaps even to a certain extent in all these cases) there is a combination of top-down and bottom-up justificatory procedures being employed together.

McMullin describe will typically only work when the target system is related to the model system "smoothly" via an idealization. For those models that represent their target systems via some relation other than idealization, such an approach will not typically work.[21] On my view, this justificatory step is important insofar as it plays a role in distinguishing those models that are merely phenomenological, "saving the phenomena," from those models that are genuinely explanatory.

6.5 From structural explanations to structural model explanations

With this general account of model explanations in place, we are now in a position to begin to see how it is that classical structures can explain quantum phenomena. Applying this framework to the case of diamagnetic Rydberg atoms presented in Section 5.4, I argue that the closed classical trajectories are able to genuinely explain the resonances in the quantum spectrum above the ionization limit, insofar as the behavior of a *classical* electron moving under the joint action of the Coulomb and magnetic fields can be used to model certain structural features of the quantum spectrum (e.g., the positions of the absorption peaks in the time domain). There is a pattern of counterfactual dependence in that one can say precisely how the quantum absorption spectrum would have been different if the classical closed orbits had been changed. One is justified in saying that the closed orbits *explain* these resonances because one can rigorously show that it is these same classical closed orbits that are picked out through constructive interference of the outgoing and returning quantum electron wave; in other words, the resonances to be explained fall within the domain of applicability of the classical closed orbit model.

Similarly, in the quantum phenomenon of scarring, discussed in Section 5.5, one can say that the behavior of the classical trajectories in the classical stadium billiard can be used to model the behavior of the wavefunctions in the quantum stadium billiard. The periodic classical trajectories (and their Lyapunov exponents) are able to genuinely explain the morphology of the scarred wavefunctions by providing part of a model explanation for that phenomenon. More specifically, there is a pattern of counterfactual dependence of the quantum phenomena on these classical dynamical structures. Furthermore, this pattern of dependence is able to support Woodward's counterfactual w-questions. That is, this pattern of dependence allows one to say *precisely* how the quantum wavefunction morphology *would* change if, for example, the classical periodic orbit had been different, or if the Lyapunov exponent of that same orbit had taken on another value.

[21] As I will argue in the next section, classical trajectories are not properly thought of simply as an idealization of the quantum dynamics.

In addition to exhibiting a pattern of counterfactual dependence, I have argued that in order for a model to be genuinely explanatory (and not just a phenomenological model), there must be a further justificatory step, establishing the legitimacy of the model, as a model of that domain of phenomena in the world. As I described earlier, this justificatory step usually proceeds in one of two ways: either bottom up through something like a de-idealization analysis, or top down by way of an overarching theory. In the case of wavefunction scarring, for example, this justificatory step cannot proceed via a de-idealization analysis. This is because the classical trajectories are not properly thought of as an "idealization" of the quantum dynamics. In other words, one does not asymptotically recover the quantum wavefunctions by de-idealizing and "adding something back in" to the classical trajectories. Instead this justificatory step proceeds top down, by way of Gutzwiller's periodic orbit theory, which specifies precisely how the classical trajectories can be properly used to model certain features of the quantum dynamics.

Even though I have now established that these classical dynamical structures meet the conditions for providing a model explanation of the quantum phenomena, there still remains the question of what *kind* of model explanation they provide. The semiclassical explanations of quantum phenomena (such as scarring and the anomalous oscillations in the spectrum) do not seem to fall into any of the three types of model explanations discussed so far. First, they fail to be (constitutive) mechanistic model explanations, because the classical trajectories are not properly thought of mechanistically as the "parts" of the wavefunction. Second, these examples of semiclassical explanations also fail to be a case of covering-law model explanations, since the explanans do not essentially cite a law of nature, nor are the explanans entirely true. Unlike Elgin and Sober's example, it is not the case that the fictions in the model are irrelevant to the explanation. In their case, it is the true parts of the model that do the explanatory work, and the point of the "harmless" analysis is to show that the fictions do not get in the way of this explanation. In the present case, however, it is the fictions themselves – i.e., the classical trajectories that do not, strictly speaking, exist in quantum systems – that do the explanatory work.

Finally, I think it is also a mistake to construe semiclassical explanations as causal model explanations. In the case of wavefunction scarring, for example, although it is tempting to use causal language when describing this phenomenon, there is no sense in which the classical trajectories can be said to *cause* the wavefunction scarring. Making use of Woodward's account of causal explanation, we can see that, while there is a clear pattern of counterfactual dependence, it does not make sense to construe this dependence as a set of possible interventions or manipulations. That is, it does not make sense physically to talk about intervening in the classical trajectories to change the quantum wavefunctions. Rather, one intervenes in or manipulates the physical system in some way, such as by changing the shape of the billiard, and then both the classical trajectories and the quantum wavefunction morphologies change.

In his book on causal explanations, Woodward very briefly discusses these sorts of cases in which there is a pattern of counterfactual dependence, and yet one cannot give it an interventionist interpretation. The example Woodward gives (due to Richard Healey) is that of explaining the stability of planetary orbits by appealing to the number of dimensions of space-time. This example, he writes,

> fits quite well with the idea that explanations provide answers to what-if-things-had-been-different questions on one natural interpretation: we may think of the derivation as telling us what would happen if space-time were five-dimensional, and so on. On the other hand, it seems implausible to interpret such derivations as telling us what will happen under *interventions* on the dimensionality of space-time.
>
> *(Woodward 2003, p. 220; see also Hitchcock and Woodward 2003, p. 191)*

Woodward simply notes that this case indicates that there is some sort of noncausal explanation, though he does not go on to specify what kind of noncausal explanation it is.

The example he cites is, I think, indicative of a type of explanation that one finds quite frequently in mechanics and fundamental physics. Following Peter Railton (1980, Section II.7) and R.I.G. Hughes (1989), I shall refer to such explanations as *structural explanations*. Moreover, I shall argue that this is the type of explanation that is involved in classical trajectories explaining quantum phenomena.

Very broadly, a structural explanation is one in which the explanandum is explained by showing how the (typically mathematical) structure of the theory itself limits what sorts of objects, properties, states, or behaviors are admissible within the framework of that theory, and then showing that the explanandum is in fact a consequence of that structure.[22] Hughes gives the following nice example of a structural explanation:

> Suppose we were asked to explain why, according to the Special Theory of Relativity (STR), there is one velocity which is invariant across all inertial frames ... A structural explanation of the invariance would display ... the admissible coordinate systems for space-time that STR allows ... [and] then show that there were pairs of events, e_1 and e_2, such that, under all admissible transformations of coordinates, their spatial separation X bore a constant ratio to their temporal separation T, and hence the velocity X/T of anything moving from e_1 and e_2 would be the same in all coordinate systems. It would also show that only when this ratio had a particular value (call it "c") was it invariant under these transformations.
>
> *(Hughes 1989, pp. 198–9)*

[22] This definition of structural explanation is my own, and is not exactly identical to the definitions given by other defenders of structural explanations, such as Railton, Hughes, and Clifton. Nonetheless I think this definition better describes the concrete examples of structural explanations that these philosophers give, and is preferable given the notion of "model" being used here.

That is, in this structural explanation, one explains the invariance of the speed of light by showing that it is a consequence of the structure of special relativity itself (and the space-time that it postulates), given the restrictions this structure imposes on the admissible coordinate systems and transformations in the theory. As Hughes notes, "[t]his kind of explanation would not fit the D-N model in any obvious way; nor is it a causal explanation: no 'shrinking rods' or 'slowing down of clocks' is appealed to … It is a purely structural explanation" (Hughes 1989, p. 199).

More recently Rob Clifton (1998) has picked up on Hughes's account of structural explanation, and argued that one frequently finds this form of explanation in quantum theory. Like Hughes, he notes that "[i]t is natural to call explanations based on this maxim *structural* to emphasize that they need not be underpinned by causal stories and may make essential reference to purely mathematical structures that display the similarity and connections between phenomena" (Clifton 1998, p. 7). Following these thinkers, I will take structural explanations to be a distinctive form of explanation that does not fall under either the covering law or causal accounts.[23]

With this characterization of structural explanations in hand, we are now in a position to answer the question of what kind of model explanation these classical dynamical structures provide: they provide a *structural model explanation*. A structural model explanation, then, is one in which not only does the explanandum exhibit a pattern of dependence on the elements of the model cited in the explanans, but in addition, this dependence is a consequence of the structural features of the theory (or theories) employed in the model.

This account of structural model explanation can be contrasted with the other versions of model explanations, discussed earlier, in which the dependence of the explanandum on the elements of the model is construed as either the elements of the model *causally producing* the explanandum (in the case of causal model explanations), the elements of the model *being the mechanistic parts which make up* the explanandum-system whole (in the case of mechanistic model explanations), or the explanandum being a *consequence of the laws* cited in the model (in the case of covering-law model explanations). In other words, to answer the question of what kind of model explanation it is, one needs to articulate what we might call the "origin" of this counterfactual dependence. In the case of a structural model explanation, this counterfactual dependence is understood as arising from the structural (or mathematical) features of the theories being employed in the model.

In conclusion, classical trajectories – even though they are fictions – can genuinely explain the morphology of quantum wavefunctions and the anomalous sequence of peaks in the absorption spectrum of diamagnetic Rydberg atoms. They do so by providing part of a structural model explanation of these phenomena.

[23] A more complete characterization and discussion of structural explanations shall be given in a future work.

As the images of wavefunction scarring vividly show, these quantum wavefunctions exhibit a pattern of dependence on the classical trajectories and their stability (Lyapunov) exponents. We can understand this counterfactual dependence as arising, not from the trajectories in any way *causing* the wavefunction morphologies, but rather from the structural features of periodic orbit theory, which specify precisely how these classical and quantum structures are to be linked. I have argued that it is both (a) the existence of periodic orbit theory, which can be used to justify the modeling of quantum phenomena with classical structures, and (b) the wide range of what-if-things-had-been-different questions that these classical structures can answer, that indicate that they are not simply providing a phenomenological model, which "saves the quantum phenomena." Rather these classical structures are yielding genuine physical insight into the nature of the quantum dynamics – a physical insight that, I shall argue next, is missing from a purely quantum-mechanical deduction of the phenomenon of wavefunction scarring from the Schrödinger equation.

6.6 Putting understanding back into explanation

Once the legitimacy of appealing to fictional classical trajectories in explaining wavefunction scarring has been established, the next question arises: Why would one bother to do so? My claim is not that there is no purely quantum-mechanical explanation for this phenomenon, but rather that such an explanation, that omits reference to classical periodic orbits, is deficient. Although one can "deduce" the phenomenon of wavefunction scarring by numerically solving the Schrödinger equation, such an explanation fails to provide adequate *understanding* of this phenomenon.

There is a tradition in the philosophy of science that argues that the notion of understanding has no legitimate place in a rigorous account of scientific explanation. Hempel, for example, writes, "[E]xpressions such as 'realm of understanding' and 'comprehensible' do not belong to the vocabulary of [the logic of explanation], for they refer to the psychological or pragmatic aspects" (Hempel 1965, p. 413).[24] His concern is that the notion of understanding is necessarily relative and subjective, depending on "P's beliefs at the time as well as his intelligence, his critical standards, his personal idiosyncrasies, and so forth" (Hempel 1965, p. 426). However, subsequent philosophers, such as Michael Friedman (1974), have cogently argued that one can have a perfectly objective notion of scientific understanding, that does not vary from individual to individual. Indeed to leave the

[24] More recently J.D. Trout (2002) has similarly defended the view that it is a mistake to include the notion of understanding in an account of scientific explanation.

notion of understanding out of a philosophical account of explanation misses one of the primary functions of scientific explanation.

There are many different dimensions to our everyday notion of understanding. Wesley Salmon, for example, writes,

> Our understanding is enhanced (1) when we obtain knowledge of the hidden mechanisms, causal or other, that produce the phenomena we seek to explain, (2) when our knowledge of the world is so organized that we can comprehend what we know under a smaller number of assumptions than previously, and (3) when we supply missing bits of descriptive knowledge that answer why-questions and remove us from the particular sorts of intellectual predicaments. Which of these is *the* function of scientific explanation? None *uniquely* qualifies.
> *(Salmon 1989, pp. 134–5)*

Although some have taken the notion of understanding to be irreducibly psychological and subjective, I think there is an objective, non-psychological notion of understanding that can be straightforwardly incorporated into an account of scientific explanation, and that is in terms of what Hitchcock and Woodward call "explanatory depth."

On their account, not only can there be more than one scientific explanation for a given explanandum-phenomenon, but it is also the case that some of these explanations are deeper than others. They argue, "One generalization can provide a deeper explanation than another if it provides the resources for answering a greater range of what-if-things-had-been-different questions … That is, generalizations provide deeper explanations when they are *more general*" (Hitchcock and Woodward 2003, p. 198). They emphasize that rival accounts of explanation (such as Hempel's D–N model) have been unable to provide an adequate account of explanatory depth because they have construed "more general" in the wrong way. That is, they have taken generality to mean that the explanation should apply to different sorts of systems and objects than the one under consideration. By contrast, Hitchcock and Woodward argue that "[t]he right sort of generality is rather generality with respect to *other possible properties of the very object or system that is the focus of explanation*" (Hitchcock and Woodward 2003, p. 182).

I think this account of explanatory depth in terms of the ability to answer a wider range of w-questions (that is, what-if-things-had-been-different questions), is largely correct. Very roughly, then, we can take explanatory depth to be a measure of *how much information* the explanans provides about the system of interest. As Hitchcock and Woodward rightly emphasize, explanatory depth is not one dimensional – there are different dimensions along which explanatory depth can be measured, and moreover, these various dimensions can compete with each other. One can begin to get a sense of the various dimensions along which explanatory depth can be measured, by thinking about the various *classes* of w-questions one might ask about the target system.

The sort of examples that Hitchcock and Woodward discuss are typically examples in which the deeper explanation is afforded by the more *fundamental* theory, such as in their example that general relativity will provide deeper explanations than Newtonian gravitational theory (Hitchcock and Woodward 2003, p. 184). If, however, we adopt their account of explanatory depth as being able to answer a range of w-questions, and further recognize that there are different *kinds* of w-questions one can ask about the explanandum system, then, I argue, there can be situations in which *less* fundamental theories can provide deeper explanations than more fundamental theories.

In the case of diamagnetic Rydberg atoms, the purely quantum-mechanical explanation of the resonances above the ionization limit involves deducing the spectrum from the Schrödinger equation. Although a hydrogen atom in a strong magnetic field is prima facie one of the simplest quantum systems, it is highly nonseparable, making the quantum calculations exceedingly difficult, and even more so in the high-energy regime where the anomalous resonances occur. As Kleppner and Delos explain, "Brute force quantum solutions based on diagonalizing larger and larger basis sets rapidly ran out of power as the energy was raised, and for the positive energy regime [i.e., above the ionization limit] the method totally failed" (Kleppner and Delos 2001, p. 596). It was not until 1991 that Delande and Gay at the École Normale Supérieure were able to develop an approximation method coupled with a numerical simulation algorithm whereby the spectrum of a diamagnetic hydrogen atom could be reliably produced theoretically. Although this was an important and remarkable theoretical achievement, "the quantum solution to the diamagnetic hydrogen problem is austere – accurate but apparently indifferent to the underlying physical processes" (Kleppner and Delos 2001, p. 596). In other words, although one can, with great difficulty, construct a quantum D–N explanation of the anomalous spectra, such an explanation offers virtually no understanding of the physics underlying this phenomenon.

Similarly, the complexity of the high-energy eigenstates of the quantum stadium billiard also means that the Schrödinger equation for this system can only be solved numerically. Simply showing, as a purely quantum-mechanical D–N explanation does, that certain eigenstate morphologies follow from the dynamical law with the relevant boundary conditions for this system, once again gives little physical insight into this phenomenon. My claim is not that such quantum-mechanical explanations are false or eliminable. Indeed if one could not give a purely quantum-mechanical explanation of either the resonances or scarring, this would raise concerns about the status of quantum mechanics as a fundamental theory. Rather, the sense in which such quantum-mechanical explanations are deficient is that they fail to provide adequate physical insight into, or understanding of, these phenomena.

Indeed this is precisely the sort of justification that physicists give for using the semiclassical models. For example, regarding the diamagnetic hydrogen atom, the physicist Dieter Wintgen and his colleagues write,

> The high dimensionality of the problem combined with the vast density of states can make the [full quantum-mechanical] calculations cumbersome and elaborate … Furthermore, the problem of understanding the structure of the quantum solutions still remains after solving the Schrödinger equation. Again, simple interpretation of classical and semiclassical methods assists in illuminating the structure of solutions.
>
> *(Wintgen* et al. *1992, p. 19)*

As they explain, the advantage of semiclassical approaches is not simply that the calculations are easier to carry out. More importantly, it is that the semiclassical models provide *more information* about the structure of the quantum dynamics than do the full quantum calculations. That is, the semiclassical model allows one to answer a wider variety of w-questions, about how the system would behave if certain parameters were changed, and provides this information without having to explicitly carry out the tedious quantum calculations for each possible case. By providing an underlying picture of the quantum dynamics, these semiclassical models can reveal important structural features of the quantum dynamics that would otherwise remain obscured. Hence if one is trying to explain a structural phenomenon such as wavefunction scarring or the intrashell resonances of the helium atom (Section 5.3), then it is the semiclassical model explanations that turn out to be deeper than the quantum explanations, insofar as the semiclassical models allow you to answer more w-questions about the system of interest.

Classical structures, such as closed and periodic orbits, provide a level of understanding of these phenomena that the purely quantum-mechanical explanations do not. By relating the quantum system to these classical structures, one is able to make use of the existing knowledge and resources of the classical models and theories in furthering our understanding of the quantum system.[25] The deeper understanding that these classical structures provide is further evidenced by the fact they enable one to make a variety of successful predictions about the behavior of the quantum system.[26] There is thus a certain symmetry insofar as these model explanations can also be used for predictions.[27] Without knowledge of the classical orbits, our understanding of the quantum spectra and wavefunction morphologies is incomplete. In sum, although reference to classical structures is in some sense eliminable

[25] This move is not simply an explanation by "reduction to the familiar," since classically chaotic models, which are themselves still the subject of difficult mathematical and scientific investigation, are far from familiar in the usual sense of the term.

[26] For examples of these predictions in the case of scarring see Kaplan (1999).

[27] In the context of D–N explanations, Hempel refers to this sort of symmetry as the "thesis of structural identity," whereby every adequate explanation is potentially a prediction and every adequate prediction is potentially an explanation (Hempel 1965, p. 367).

from the explanation of phenomena such as diamagnetic Rydberg spectra and wavefunction scarring, such an elimination comes at the rather high cost of understanding.

As we shall see next, the semiclassical research described in the last chapter not only poses a challenge for our traditional accounts of scientific explanation, but also calls into question the adequacy of current philosophical frameworks for thinking about intertheory relations. More specifically, I shall argue that an adequate account of the relation between classical and quantum mechanics requires recognizing the many structural continuities and correspondences pervading these theories. Building on the views of Paul Dirac, I shall outline a new approach to intertheory relations that I call interstructuralism. I shall show that interstructuralism not only provides a better account of the relation between classical and quantum mechanics but also sheds new light on current debates over structural realism.

7

A structural approach to intertheoretic relations

There are more things in heaven and earth, Horatio, Than are dreamt of in your philosophy.

Shakespeare, Hamlet, *Act 1 Scene 5*

7.1 The challenge of semiclassical mechanics: Reassessing the views

Semiclassical mechanics significantly broadens our understanding of the relationship between classical and quantum mechanics. It reveals that there is much more to the relation between these theories than simply showing that classical behavior can be recovered from quantum systems in the limit of large actions ($\hbar \to 0$) or large quantum numbers ($n \to \infty$).[1] Although these limits are an important piece of the puzzle, the usual reductionist account of the quantum–classical relation in terms of these limits alone does not tell the whole story.

More specifically, research in semiclassical mechanics has uncovered a number of surprising theoretical and experimental *correspondences*. Indeed the program of semiclassical research can be broadly understood as the theoretical and experimental task of uncovering these correspondences and extending them in new ways. On the theoretical side, semiclassical methods such as EBK quantization, Gutzwiller's trace formula, and closed orbit theory show how classical structures can be used to construct quantum quantities. These semiclassical methods have proven to be not just useful calculational tools, but also important investigative tools, in that they are often able to provide physical insight into the structure of the quantum dynamics, and hence lead to new discoveries. The success of semiclassical methods in providing both calculational and theoretical insight suggests that there is a greater continuity of dynamical structure across these theories than is traditionally recognized.

[1] As we saw in Chapter 1, even this is far from a trivial or fully accomplished task.

Recent experimental work in semiclassical mechanics provides even stronger evidence for this structural continuity between classical and quantum mechanics. As we saw in Chapter 5, semiclassical experiments are revealing many surprising ways in which classical structures are manifesting themselves in quantum phenomena. For example, the research on the behavior of Rydberg atoms in strong external fields, described in Section 5.4, shows how classical closed orbits are manifesting themselves in the quantum absorption spectrum. It is worth emphasizing again that the results of these experiments were *not* to show that the Rydberg electron "mimics" the classical electron path in the classical limit, as one might expect on the usual reductionist view of intertheory relations. Rather, these experiments show that there are striking correspondences between specific classical and quantum structures – correspondences that are furthermore determinable directly from the experimental data.[2] This suggests that these correspondences are not just some mathematical trick of the theoretician, but rather are correspondences that are structuring the empirical phenomena themselves.[3]

The many correspondences revealed by semiclassical research challenge the adequacy of our current philosophical accounts of intertheory relations. One of the central theses of this book is that the recognition of such correspondences needs to be an essential part of any adequate account of the relationship between classical and quantum mechanics. It is far from clear, however, that the traditional reductionist view (described in Chapter 1 as Nickel's reductionism$_2$) can make adequate sense of the structural continuity we find between classical and quantum mechanics. Against this reductionist view, we have seen that there are interesting and important things to say about the relation between these theories other than simply showing that one can recover the classical predictions from the quantum equations in some limit. The requirement that the classical equations emerge from the quantum equations in the limit of some parameter (or combination of parameters) in no way necessitates or explains these correspondences.

In searching for a more adequate philosophical account of intertheory relations, I introduced (in Chapters 2, 3, and 4) the philosophical views of three of the founders of quantum theory: Werner Heisenberg, Paul Dirac, and Niels Bohr. As we saw, none of these physicists adopted the usual reductionist or eliminativist accounts of the quantum–classical relation. Instead, all three accorded to classical mechanics a role of continued *theoretical* importance; none of them took classical mechanics

[2] Recall that Haggerty *et al.* (1998) were able to measure the (fictional) classical trajectories (that is, the electron position as a function of time) *directly* from the quantum spectroscopic data. These classical trajectories are taken to be important for explaining phenomena such as the Stark spectrum of Rydberg lithium. An account of how these fictional classical structures can be explanatory was given in Chapter 6.

[3] At this point I wish to remain neutral about the realism–antirealism question. The observation of these empirical correspondences is at least prima facie compatible with Bas van Fraassen's antirealist constructive empiricism, for example, which shall be discussed below.

to be a discarded theory, rendered obsolete for all but "engineering" purposes by the new quantum theory. Although they all shared this commitment to the on-going theoretical importance of classical mechanics, the specific roles that they accorded to this theory were quite different. This led each of them to develop a very different philosophical account of intertheory relations. The question I want to examine now is whether any of these three frameworks can provide the basis for a more adequate philosophical account of the relationship between classical and quantum mechanics – one that can make sense of the recent insights from semi-classical research.

As we saw in Chapter 2, Heisenberg denies that quantum mechanics is a universal theory, arguing instead that there is some domain of phenomena for which classical mechanics is the perfectly accurate and final description. His account of closed theories was shown to be a realist form of theoretical pluralism, according to which nature is governed in different domains by different systems of laws. Heisenberg further argues that quantum mechanics had to make a clean break from classical mechanics in its development, and that the transition between these theories was a revolutionary and discontinuous one. As we saw in Section 2.3, he even takes the concepts of classical and quantum mechanics to be incommensurable with one another, in a manner quite similar to Thomas Kuhn. Moreover, Heisenberg viewed quantum mechanics as a closed axiomatic system that is complete in itself, and hence in no way dependent on other theories or concepts for its application.[4]

It is evident even in Heisenberg's own 1925 paper, however, that quantum mechanics did not in fact make a clean break, and instead was built on the backbone of classical mechanics.[5] The recent theoretical and experimental work in semiclassical mechanics shows that these classical bones are still manifesting themselves in the quantum phenomena. Moreover, the hybrid methods characteristic of semiclassical research show that, *pace* Heisenberg and Kuhn, quantum and classical concepts are *not* incommensurable and can in fact be combined in empirically adequate and conceptually fruitful ways. Given Heisenberg's realist construal of classical and quantum mechanics as distinct closed theory domains, it is not clear that he would recognize a field such as semiclassical mechanics for which both classical and quantum descriptions are required. Indeed by interpreting quantum mechanics as a closed theory that is complete in itself, he rules out the possibility that there may be some quantum phenomena that depend on classical mechanics for their full explanation.[6] By advocating a strong form of theoretical pluralism, Heisenberg has left

[4] I argued in Section 2.6 that Heisenberg's occasional public endorsements of Bohr's indispensability of classical concepts is in fact incompatible with his considered views on closed theories. See also Beller (1996, p. 184).

[5] Heisenberg's 1925 paper was discussed at the end of Section 4.2, and this point was also defended in connection with Dirac's work in Section 3.2.

[6] Recall the defense of the view that classical structures are playing an indispensable role in explaining quantum phenomena given in Section 6.6.

little room for making sense of the thorough-going correspondences between the dynamical structures of classical and quantum mechanics.

By contrast, there are a number of affinities between Bohr's account of inter-theoretic relations and this recent work in semiclassical mechanics. First, both Bohr and the semiclassical theorists emphasize the theoretical continuity between classical mechanics, the old quantum theory, and the new quantum mechanics. Recall that Bohr viewed quantum theory as a rational generalization of the classical theories, and the correspondence principle that he discovered is the link that ties all three of these theories together. Similarly, semiclassical mechanics is not only taken to be an extension of Bohr's old quantum theory, but also to have a firm grounding as an approximation to the full quantum theory. As such, it provides a strong link between all three of these theories, in a manner consistent with Bohr's emphasis on continuity.

Second, Bohr's correspondence principle can be understood as an example of precisely the sort of structural continuities between classical and quantum mechanics emphasized in semiclassical research. As I argued in detail in Section 4.2, Bohr's correspondence principle is not the asymptotic agreement of quantum and classical prediction for large quantum numbers. Rather, it is the recognition that there is a law-like connection between the allowed quantum jumps between stationary states in an atom and the harmonics present in the classical description of the electron's motion. Another way of describing Bohr's correspondence principle is that, despite the very different pictures behind the classical and quantum accounts of radiation, it turns out that the number τ that in classical mechanics labels the harmonic components of the Fourier decomposition of the periodic classical trajectory *equals* the number τ in the old quantum theory that describes the difference in stationary states jumped. Moreover, as we saw in Heisenberg's 1925 paper, this correspondence principle rule gets incorporated into modern quantum mechanics insofar as this number τ *also equals* the difference between the indices of Heisenberg's matrix components.[7] So, for example, if there is no third harmonic in the classical motion, then quantum jumps of three stationary states are not allowed in the old quantum theory, and those matrix elements whose indices have a difference of three will be zero in modern quantum mechanics. This is a remarkable structural continuity between what are prima facie quite different theories. Although Bohr also recognized that a number of statistical asymptotic agreements (between frequencies, intensities, etc.) follow from this, it is the structural correspondence between

[7] Once again, I am by *no* means claiming that all of quantum mechanics is just a formalization of the correspondence principle. Quantum mechanics of course encompasses a much wider scope of states and processes than those to which the correspondence principle applies.

classical and quantum mechanics – not any asymptotic agreement – that is fundamental.[8] The lesson I want to draw here is that when someone asks, "What is the relation between classical and quantum mechanics?", these sorts of deep structural continuities must be an important part of any adequate answer.

In an insightful paper, Robert Batterman (1991) explicitly connects up Bohr's correspondence principle with modern semiclassical methods. Although his interpretation of Bohr's correspondence principle differs slightly from the one I have given here, the spirit is very much the same. Batterman defines Bohr's correspondence principle as follows: "Thus the CP is a statement to the effect that the radiative processes to which the second postulate [$h\nu = E' - E''$] applies are 'correlated' with, or 'correspond' to, mechanical vibrations or periodic motions of the charged particles" (Batterman 1991, p. 203). He rightly recognizes that the correspondence principle is not the (statistical) asymptotic agreement of predictions, and moreover that it is something that *explains* or *justifies* that asymptotic agreement.

Batterman's argument in this paper is that, if we interpret the correspondence principle more generally as a correspondence between classical periodic motions and discrete quantum properties, then not only can the EBK "torus quantization" method (introduced in Section 5.2) be understood as an extension of Bohr's correspondence principle, but even Gutzwiller's trace formula can be seen as a vindication of Bohr's correspondence principle. The trace formula, recall, relates classical periodic orbits to the quantum density of states – even when the classical motion is chaotic. Batterman concludes that the problem of quantum chaos, understood as the absence of chaotic behavior in quantum systems despite its pervasiveness in classical systems, does not in fact threaten Bohr's correspondence principle, when this principle is properly understood.

While I think this perhaps stretches the meaning of Bohr's correspondence principle a bit too far, I do think Batterman is right to see the sort of correspondences identified by modern semiclassicists as being in the same tradition as Bohr's correspondence principle. Rather than viewing Gutzwiller's periodic orbit sum as an instance of Bohr's correspondence principle, what we should perhaps say instead is that Martin Gutzwiller has discovered yet *another* correspondence principle. And perhaps Einstein, Brillouin, and Keller discovered a third. Indeed, as we saw in Section 5.4, this is the way that John Delos understands his discovery of the relation between the oscillations in the diamagnetic Rydberg spectrum and the closed classical trajectories, which he refers to as the discovery of the *correspondence*

[8] Although I do not have a proof of this, my intuition is that one could cook up a way of having a statistical asymptotic agreement between classical and quantum mechanics (the frequencies, for example), without having this law-like correspondence between the classical and quantum structures. Hence to emphasize the former rather than the latter is to completely miss this important insight of Bohr's.

principle for finite resolution spectra (Delos and Du 1988, p. 1448).[9] In this sense, semiclassical mechanics is very much in Bohr's tradition of uncovering the theoretical and experimental structural continuities or correspondences between the quantum and classical theories.

More generally, both Bohr and the semiclassical approach accord to classical mechanics a central role in furthering our understanding of quantum theory. Despite these important affinities, however, there are limitations to applying Bohr's philosophical account of intertheory relations to semiclassical research. These limitations stem primarily from Bohr's later, more entrenched, views about complementarity and the indispensability of classical concepts for modern quantum mechanics. For Bohr, classical concepts are required to give the quantum formalism its meaning, and they are necessary for communicating the results of any experiment.[10] For the semiclassical theorist, by contrast, there is no such requirement for the language of science. Instead, as we have seen, the value that classical mechanics brings to quantum systems is simply that of providing further physical insight into the structure of the quantum dynamics.

Moreover, it is far from clear that Bohr would embrace the use of fictional classical trajectories in the quantum domain, insofar as these trajectories would seem to violate his principle of complementarity, which restricts the simultaneous application of concepts such as position and momentum. More to the point however, for Bohr, classical mechanics is primarily an *interpretive* tool for quantum theory, and simply does not play the broader calculational, investigative, and explanatory roles that it does in semiclassical research. Although Bohr's account of the relation between classical and quantum mechanics provides an important alternative to the usual philosophical frameworks for thinking about intertheoretic relations, in the end I do not think it is the most helpful one for making sense of the fertility of semiclassical research.

The third alternative view of intertheoretic relations (introduced in Chapter 3) is Dirac's structural approach. Like Heisenberg and Bohr, Dirac accords to classical mechanics a role of continued theoretical importance. Unlike in Heisenberg, however, this role is not as the final description of some distinct domain, and unlike in Bohr, it is not as some sort of requirement for the language of science. Rather, Dirac sees classical mechanics as the basis of a formal analogy from which quantum mechanics was – and is continuing to be – developed. There is thus a sense in which Dirac, like Bohr, views quantum mechanics as a rational generalization of classical mechanics; for Dirac, however, it is the generalization of classical mechanics to a

[9] Although I think it is conceptually helpful to think of these as distinct correspondence principles, there are important and deep mathematical connections between these various correspondence principles, which have not been fully explicated here.

[10] This view of Bohr's was discussed in Section 4.4.

noncommutative algebra, while for Bohr it is a generalization in accordance with the correspondence principle and complementarity.

According to Dirac, the relation between classical and quantum mechanics lies not in the limiting agreement as $\hbar \to 0$ (reductionism$_2$), but rather in structural correspondences such as the one he discovered between the classical Poisson bracket and the quantum commutator (a relation that is sometimes referred to as *Dirac's correspondence principle*). A further example of this structural continuity was seen in Dirac's development of a quantum Lagrangian mechanics. What is distinctive about Dirac's approach to intertheory relations is that he takes these structural continuities to *pervade* these theories – not simply to be continuities that emerge in the classical limit. As I noted in the case of Bohr's correspondence principle, although these structural continuities can lead to an asymptotic agreement of predictions, simply having an asymptotic agreement does not guarantee this structural continuity. It is the structural continuity which is fundamental and explains the asymptotic agreement, not the other way around.[11]

Not only does Dirac take there to be a deep structural continuity across classical and quantum mechanics, but he sees in this continuity an important methodology for the further development of these theories. This methodology, which can be understood most broadly as one of analogy extension, in many cases takes the more precise form that I called the *reciprocal correspondence principle methodology*.[12] In Section 3.2, I defined this methodology as the use of problems in quantum theory to guide the further development of classical mechanics so that those results obtained in the classical context can then be transferred back to aid in the further development of quantum theory. For Dirac, both classical and quantum mechanics are open theories, and the analogies and correspondences between these theories can, through new developments in mathematics and physics, be extended in new ways. Thus, although Dirac is a firm believer in the unity of science, this unity rests not on a traditional reductive relationship, but rather on the thoroughgoing structural analogies and correspondences between these theories, which are continuing to be discovered and extended.

Dirac's structural continuity view of the relation between classical and quantum mechanics provides the outline for a new, more adequate approach to intertheory relations – one capable of incorporating the recent insights from semiclassical

[11] In Dirac's 1925 paper we can see him reasoning in a similar way, arguing that it is the fact that the quantum and classical equations "obey the same laws" that is important, not the weaker claim that the classical equations are recovered from the quantum equations as $h \to 0$ (p. 649).

[12] As I noted in Chapter 3, using the term "principle" is somewhat misleading in this context, since it designates a methodology not a principle. As we saw in Chapter 4, Bohr's correspondence principle really is a principle – a law-like relation – not just a heuristic methodology. Nonetheless, in the case of both Dirac and Bohr, there are genuine correspondence principles underlying these heuristic methodologies, which in some sense make them possible.

research. There are a number of striking affinities between Dirac's approach and semiclassical research. First, the recent explosion of work in chaos theory shows that classical mechanics is indeed an open theory; far from being discarded or closed, classical mechanics is yielding new theoretical insights, and is continuing to be developed and extended in new ways. These recent developments in classical mechanics have in turn led to new developments in quantum mechanics, such as those described in the field of quantum chaos.

The second respect in which Dirac's approach to intertheory relations provides a natural framework for accounting for semiclassical research lies precisely in the emphasis that both approaches place on these structural correspondences between classical and quantum mechanics. As we saw earlier, both Dirac and the semiclassical theorist are interested in identifying the quantum analogs of classical structures. Furthermore, both of them are interested in showing how these formal analogies can be extended and made more precise through the development of new mathematical methods. By conceiving of the relation between classical and quantum mechanics broadly in terms of these structural analogies, rather than the more narrow conception of a limiting agreement of predictions, Dirac's approach to intertheory relations is better able to accommodate the wide-ranging theoretical and experimental correspondences revealed by semiclassical research. Moreover, it is the dynamic nature of these analogies and correspondences, which Dirac recognizes, that allows classical mechanics to play the much broader calculational, investigative, and explanatory roles that are characteristic of semiclassical mechanics.

Third, semiclassical methods can be broadly understood as part of Dirac's method of analogy extension. More specifically, semiclassical formulas, such as Gutzwiller's trace formula (periodic orbit theory), are the sort of two-way street between classical and quantum mechanics envisaged by Dirac's reciprocal correspondence methodology. That is, not only can one use Gutzwiller's trace formula to calculate the quantum density of states from the classical periodic orbits, but one can also use "Gutzwiller's correspondence in reverse" to calculate the classical periodic orbits from knowledge of the quantum system. This latter inverse correspondence principle has proven to be particularly useful in chemistry, as Gregory Ezra describes: "Although the semiclassical po [periodic orbit] expressions for the quantum level density can in principle be used in the 'forward' mode to compute semiclassical eigenvalues for individual levels … [i]n the work described below, we use po theory 'in reverse' to extract information on classical phase space structure directly from vibrational eigenvalues" (Ezra 1998, p. 41). He further explains why such an approach is so useful in chemistry:

The fundamental aim of molecular spectroscopy is to derive information about the a priori unknown molecular Hamiltonian $\hat{H}$ and its associated dynamics … Our understanding of the quantum dynamics of $\hat{H}$ is however greatly enhanced by study of the corresponding *classical* Hamiltonian H ... [C]hemists continue to picture molecules as mechanical 'ball-and-spring' objects, while the trajectories of ball bearings rolling on potential surfaces have long served to illustrate the qualitative influence of potential energy surface shape upon reactive dynamics. It is therefore important to develop methods for the analysis of molecular spectra that are (as far as possible) rooted firmly in the underlying classical mechanics of the molecular Hamiltonian; here semiclassical methods ... provide an essential bridge between the classical and quantum aspects of the problem.

(Ezra 1998, p. 38)

In other words, the correspondences embodied in semiclassical formulas not only allow for classical mechanics to be used in the solution of quantum problems, but also for quantum mechanics to be used in the solution of classical problems. Once these solutions are worked out for the classical case, this knowledge can then be transferred back to obtain a deeper understanding of the quantum system. Note the similarities between this semiclassical approach and Dirac's own description of his reciprocal correspondence principle methodology quoted in Section 3.2 (1951a, p. 20). Although the particular problems that Dirac and the semiclassical theorists are working on are quite different, the spirit of their methodology, I argue, is very much the same.

In sum, four different approaches to intertheory relations were considered here: the traditional reductionist view, Bohr's generalization thesis, Heisenberg's theoretical pluralism, and finally Dirac's structural correspondences. I argued above that Dirac's structural approach to intertheory relations provides the most natural framework for understanding recent developments in semiclassical mechanics. Dirac's philosophical views are not only helpful for making sense of semiclassical research, but, as I shall argue next, they can also shed new light on contemporary debates in philosophy over structural realism. More specifically, we shall see that this view – that there is a significant continuity of structure across what is, prima facie, a case of radical theory change – has also been at the center of debates over structural realism.

7.2 Structural realism without realism ... or antirealism

At the heart of Dirac's approach to intertheoretic relations is the notion of a structural continuity between classical and quantum mechanics. More recently, a similar emphasis on structural continuity across theory change has appeared in the philosophy of science literature in discussions of structural realism. Structural realism is a view that was first articulated in the context of philosophical debates over scientific realism. Very briefly, scientific realism is the view that our best current scientific theories are giving us an accurate picture of what the world is really like; this is opposed to antirealist views, such as instrumentalism, which take our scientific theories to be nothing more than "calculational devices," useful for making predictions, but failing to give us any insight into the way the world really is. The debate over scientific realism has largely come to an impasse, however. On the one hand defenders of realism assert what is known as the "success of science argument" – that is, if it were not the case that our scientific theories were getting at the way the world really is, then the many successes that contemporary science surely enjoys would be an unexplained "miracle."[13] On the other hand, defenders of antirealism point to another argument, known as the "pessimistic meta-induction":[14] if one looks back over the history of science, even the best scientific theories have always turned out to be false, and the same is probably true of our present theories. Moreover, as historians of science such as Kuhn have argued, these scientific revolutions have failed to provide a coherent direction of ontological development.[15] The hope of the structural realists is that by paying attention to the structural features of scientific theories, rather than the particular ontologies or theoretical claims, a path can be steered between the Scylla of the success of science argument and the Charybdis of the pessimistic meta-induction.

The structural realist position can be understood as consisting of the following two theses:

1. *Continuity thesis*: There is a continuity of structure across even revolutionary theory change.
2. *Realism thesis*: It is this structure about which we should be realists.

John Worrall, in his seminal article "Structural realism: The best of both worlds?," uses the example of the replacement of Augustin Fresnel's ether wave theory of light by James Maxwell's electromagnetism to illustrate this idea of structural continuity. He notes that Fresnel developed his equations giving the relative intensities of the reflected and refracted light beams, by conceiving of light as vibrations in a mechanical aether. Although this ontology was ultimately rejected by Maxwell's theory, Fresnel's equations are nonetheless entailed by Maxwell's equations. Regarding this case, Worrall writes,

> There was an important element of continuity in the shift … and this was much more than a simple question of carrying over the successful *empirical* content into the new theory. At the same time it was rather less than carrying over the full theoretical content or full theoretical mechanisms (even in "approximate" form) … There was a continuity or accumulation in the shift, but the continuity is one of *form* or *structure*, not of content.
>
> *(Worrall 1989, p. 117)*

[13] See, for example, Putnam (1975).
[14] See, for example, Laudan (1981).
[15] "Ontology" can be roughly understood as the study of what might be called "the furniture of the world," or a description of the basic entities or kinds of things that exist. The challenge, which Kuhn and others have pointed to, is that even our quite successful past scientific theories used to assert the existence of things we no longer believe actually exist, such as phlogiston, aether, absolute time, caloric, and so on.

In other words, what Fresnel's theory gets right is certain structural features of the optical phenomena, even though the ontology and mechanisms identified in the theory turned out to be wrong.

Worrall notes that this strong notion of structural continuity, whereby the equations of the predecessor theory are preserved intact in the successor theory, is rare; more often the equations of the predecessor theory are preserved only in some limit of the equations of the superceding theory. The central point, however, is that even in cases of a radical shift in ontology, the predecessor theory was able to get certain structural features of the phenomena right (structures that are to a certain extent captured by the theory's equations) and it is this correct structural content that is preserved across the theory change.

Most of the subsequent work on structural realism has focused almost exclusively on the second thesis concerning realism. For example, James Ladyman (1998) points out that Worrall's structural realism is ambiguous between what he calls "epistemic structural realism" (ESR) and "ontic structural realism" (OSR). While ESR says roughly that all we can really know about the world is its structural aspects, OSR goes a step further and argues that the ontology of the world should itself be reconceptualized in structural terms (French and Ladyman 2003). A challenge for OSR is then to specify what such a structure could amount to, if there is no *thing* bearing that structure. Further challenges have arisen for the structural realist to specify in some non-ad hoc way, those parts of our current theories that are structural, and hence the parts that we should be realists about and can be sure will be preserved in subsequent theories (Ladyman 1998, p. 413). Without such a distinction, the realist is once again vulnerable to the pessimistic meta-induction.

In opposition to the structural realists, antirealists such as Bas van Fraassen (1997; 2006) and Otávio Bueno (1997; 1999; 2000) have defended an alternative view, which is referred to as "structural empiricism." Surprisingly, van Fraassen agrees with structural realists such as Worrall that "indeed there is a steady accumulation of knowledge in the sciences, and this knowledge deserves to be called precisely knowledge of structure" (van Fraassen 2006, p. 275). Unlike the realists, however, van Fraassen interprets this structural continuity as an accumulation of *empirical* knowledge; that is, he takes the structure in question to be nothing more than the structure of the observable phenomena, rather than an underlying, unobservable structure that should be interpreted realistically. Bueno (2000) has taken van Fraassen's empiricism a step further, denying that there need be an *accumulation* (even at the level of empirical structure) at all, and arguing instead for the weaker notion of a *partial* preservation of structure.

While there are a number of very interesting and important issues to be worked out regarding the *realism thesis* of structural realism, this debate over realism has by and large obscured the prior and more fundamental questions of whether there is in fact a continuity of structure across theory change at all, and if there is, how that continuity of structure should be characterized. In what follows, I would like to bracket the issue of realism, and instead focus on the *structural continuity thesis*.

Both structural realists, such as Steven French and Ladyman, and the structural empiricists, such as Bueno, use the formal apparatus of set theory to articulate the relevant notions of "structure" and "continuity." More broadly they situate their approaches in the context of what is known as the semantic conception of theories, according to which scientific theories are understood as families of set-theoretical models.[16] Furthermore, these scientific structuralists (on both sides of the debate) argue that the semantic conception of theories needs to be further extended by introducing the notion of a "partial structure" and a "partial isomorphism" holding between such structures.[17] They define a *partial structure* as a set-theoretic construct $A = \langle D, R_i \rangle_{i \in I}$, where D is the (non-empty) set of individuals of the domain under consideration (roughly the domain of scientific knowledge), and R_i is a family of partial relations. More specifically, a n-place partial relation R over D is understood as a triple $\langle R_1, R_2, R_3 \rangle$, where R_1 is the set of n-tuples that we know belong to R, R_2 is the set of n-tuples that we know do *not* belong to R, and R_3 is the set of n-tuples that we do not know whether or not they belong to R.[18] A *partial isomorphism* between two structures S_1 and S_2 is a partial function $f{:}D \rightarrow D'$ such that (i) f is bijective,[19] and (ii) for every x and $y \in$ D, $R_1xy \leftrightarrow R'_1 f(x)f(y)$ and $R_2xy \leftrightarrow R'_2 f(x)f(y)$.

Bueno (2000) has used this notion of a partial structure and a partial isomorphism to precisely define the notion of structural continuity at the heart of the structural realist and structural empiricist views. He writes, "In theory change (from T_1 to T_2), T_1 has only to be partially embeddable into T_2. So, for every partial model M_1 of T_1, there is a partial model M_2 of T_2 with a substructure S_2 partially isomorphic to M_1" (Bueno 2000, p. 279). Once again, this approach construes theories as families of models, and hence a continuity across theory change is understood as an ability to embed the models of the predecessor theory into the models of the successor.[20] Thus

[16] An exception to this is Worrall, whose structural realism is instead articulated within the context of the syntactic view of theories; see French and Ladyman (2003). For an introduction to the semantic conception see, for example, Suppe (1989).

[17] See, for example, Bueno, French, and Ladyman (2002), and references therein.

[18] French (1997; 2000) points out that these three *R*s can be intuitively understood in terms of Mary Hesse's "positive analogy," "negative analogy," and "neutral analogy" (Hesse 1966).

[19] Bijective means the mapping is one-to-one and onto.

[20] There is some confusion in the literature over what precisely is meant by the term "model." On the one hand, advocates of the semantic approach adopt the following notion of model: "Any structure which satisfies the axioms of a theory … is called a *model* of that theory" (van Fraassen 1980, p. 43). This is a notion of model, adopted from logic and model-theory, as a "truth-making structure." On the other hand, there is another notion of model, which is much more grounded in scientific practice, according to which a model is understood as an idealized or to some extent fictionalized representation of the phenomena (this is the sense of model employed in Chapter 6). For further disambiguation of these two senses of "model" see Frisch (2005, p. 6) or Thomson-Jones (2006), and for an attempt to synthesize these two senses of model, see French and Ladyman (1999).

in Worrall's example, structural continuity is to be understood as an embeddability of the models of Fresnel's theory into the models of Maxwell's theory. Bueno notes that the notion of isomorphism underlying the embedding relation is too strong to capture the usual sorts of continuity one finds in the history of science, and hence one must adopt the weaker notion of a partial embedding and a partial isomorphism. Hence, for Bueno, French, and colleagues, what we have in successive theories in the history of science is a *quasi-structural continuity* (Bueno 2000, p. 279).

While Bueno takes this formal account to capture the notion of structural continuity at the heart of structural realism, the lesson that he draws from this continuity is, of course, quite different. As a structural empiricist, he

> does not assign any epistemic import to the preserved structures. Continuity, in that it goes beyond empirical adequacy and informativeness, is, at best, a pragmatic factor of scientific change. In a clear sense, the empiricist does not *believe* in such structures (supposing, of course, that they go beyond empirical adequacy), but only *accepts* them.
>
> *(Bueno 2000, p. 293)*

It is perhaps surprising that both the structural realist and the structural empiricist take this formal framework to capture what they mean by structural continuity – especially since what they take that structure to be is, in the two cases, quite different. For the structural empiricists it is simply the structure of the observable phenomena, that is, the successful empirical content of the theories; whereas for the realists, it is an unobservable structure underlying, and distinct from, the empirical phenomena. It is not clear to me what advantage the empiricists gain by reformulating their position in structuralist terms, especially if, in the end, the continuity they point to is nothing other than the familiar carrying over of the successful empirical content of the theories. Indeed the point of Worrall's example of the shift from Fresnel to Maxwell was precisely to articulate a kind of continuity that was *distinct* from both empirical content and from the full ontology.

Advocates of the partial structures approach argue that this simple formal framework is able to capture an amazing variety of aspects of scientific research. They claim that the partial structures framework can not only capture the relevant (i.e., realist or empiricist) notion of structural continuity across scientific theory change, but can also account for the pattern of theory development in mathematics (Bueno 2000, Section 4). French has further argued that the partial structures framework can not only capture "the 'horizontal' inter-relationships between theories, thus providing a convenient framework for understanding theory change and construction … [but also] the 'vertical' relationships between theoretical structures and data models, accommodating, in particular the role of idealisations" (French

2000, p. 106). Furthermore, he argues that this framework can provide a unitary account of the diversity of different types of models (e.g., iconic, analog, and theoretical models) used in the sciences (French 2000; da Costa and French 2003). And finally, Bueno, French and Ladyman have argued that, if one relaxes the requirement of isomorphism, to the more general notion of *homomorphism*[21], then the partial structures approach outlined above can also capture

> two forms of applicability of mathematics: at the bottom end, as it were, the applicability of theoretical structures to what van Fraassen calls the "appearances", and at the top end, the applicability of mathematics to physics.
>
> *(Bueno, French, and Ladyman 2002, p. 498)*

In other words, the partial structures approach, by making use of partial homomorphisms is, in addition to all of these other functions, also supposed to be able to solve what Eugene Wigner called the "unreasonable effectiveness of mathematics" (Wigner 1960).[22]

The wide-reaching scope claimed for the partial structures approach begins to raise questions about the ability of this formal framework to provide any genuine insight into the problem at hand. That is, in saying that structural continuity across theory change is a partial embedding of partial structures, have we in any way really better understood the relation between classical and quantum mechanics? Do we now know how to recognize what the structural content of classical mechanics is, or which of these structures will be found preserved in quantum mechanics? My claim is not that this formal framework is wrong (which it may or may not be), but rather that this framework, with its claim to rigor, gives us a false sense of having understood and solved these substantive questions in the philosophy of science.[23] To very loosely paraphrase Einstein, while things should be made as precise as possible, they should not be made more precise.[24] In sum, while the partial structures approach has the virtues of precision and generality, it

[21] A homomorphism can be understood as a structure-preserving map from one algebraic structure (set) to another. A homomorphism is more general than an isomorphism, discussed earlier, in that an isomorphism is a homomorphism that is also one-to-one and onto.

[22] Indeed, French (2000) very cleverly titles his paper in which the partial homomorphism approach is outlined, "The reasonable effectiveness of mathematics."

[23] da Costa and French seem to be aware of this concern when they write, "At one extreme we might employ a highly developed formal approach which seeks to represent various distinctions found in scientific practice in highly technical terms. The dangers of such an approach are well-known: seduced by the scholastic angels dancing on the formal pinhead, we lose sight of the practice we are trying to understand" (da Costa and French 2000, p. S125).

[24] The original Einstein quotation is "The supreme goal of all theory is to make the irreducible basic elements as simple and as few as possible without having to surrender the adequate representation of a single datum of experience" (Einstein [1933] 1934), and is usually paraphrased as, "Things should be made as simple as possible, but no simpler."

is not clear to me that this abstract formalism has in fact given us any real purchase on the concrete debates in the philosophy of science that motivated it in the first place.

Setting aside then, for the moment, the question of how the relevant notion of "structure" is to be characterized, the question I want to examine next is whether the scientific structuralists (realist or empiricist) have offered any substantive new insight into the "continuity thesis," that is, the claim that there is structural continuity across theory change. Despite the fact that the continuity thesis is the foundation of the structuralist position – the one without which the realist thesis cannot even get off the ground – one finds surprisingly little written about the nature of this continuity. Even more surprisingly, when the continuity thesis is discussed, it is typically taken to be nothing other than the familiar Nickles reduction$_2$ relation, which describes the continuity between two theories as the equations or laws of one theory being limiting cases of the equations of the other theory. Worrall, for example, in characterizing the cumulative growth underlying structural realism writes, "The much more common pattern is that the old equations reappear as *limiting cases* of the new – that is the old and new equations are strictly inconsistent, but the new tend to the old as some quantity tends to some limit" (Worrall 1989, p. 120). Similarly Redhead, noting the challenges posed by successive theories that are related by singular limits, writes,

> We do not have continuous transformation of structure as we move away from the classical structure characterized by $1/c = 0$ in the case of relativity, and $h = 0$ in the case of quantum mechanics to the new structures with non-zero values of $1/c$ or h. Qualitatively new structures emerge, but there is a definite sense in which the new structures grow naturally, although discontinuously, out of the old structures ... So revolutions in physics, understood from the structural standpoint, can be understood progressively.
>
> *(Redhead 2001a, p. 88; see also Redhead 2001b, pp. 346–7)*

In other words, the structural continuity between theory pairs is still to be understood in terms of the limit of a parameter in the theory's equations, even in those cases where that limit is singular.[25]

The structural empiricist seems to fare no better when it comes to articulating a notion of structural continuity distinct from the familiar reductionism$_2$. van Fraassen, for example, imposes what he calls the "requirement upon succession," which he defines as the requirement that "the new theory is so related to the old that we can explain the empirical success of the old theory if we accept the new" (van Fraassen 2006, p. 275). He notes that this requirement is not a requirement of reductionism, where by reductionism van Fraassen means specifically Nagelian

[25] Recall the discussion of singular limits in Section 1.4.

reduction (that is, Nickles' reductionism$_1$). He further explains this continuity across theory change as follows:

> The relationship expressed in the above *Requirement upon succession* is not a requirement of reducibility ... Although looser, this relationship admits of rigorous proof ... The loveliest, neatest examples belong to the more mathematical sciences. So, by letting the speed of light in the Special Theory of Relativity go to infinity, you can deduce the relevant Newtonian equations ... Similarly if you let the Planck constant go toward zero in quantum mechanical deduction.
>
> *(van Fraassen 2006, p. 299)*

Despite their claimed differences, the structural realist's and structural empiricist's purportedly new accounts of "structural continuity" across theory change appear, in the end, to be none other than Nickles' familiar old reductionism$_2$ relation. One might begin to wonder: if there is not really a new form of continuity being discussed here, why introduce the potentially misleading talk of structures at all?

My answer to this prima facie rhetorical question is that while the scientific structuralists are right to point to a structural continuity across theory change, they are mistaken in trying to cash this continuity out in terms of reductionism$_2$. What the scientific structuralists seem to have missed is that the value of framing the discussion of this continuity in terms of structures lies precisely in the fact that it provides us with a *new* way of thinking about intertheory relations – one that is distinct from both Nagelian reductionism and Nickles' reductionism$_2$. What is needed, then, is a new account of intertheory relations – one appropriate to this emphasis on the structural continuities that we find between neighboring theories. I shall refer to this new account of intertheory relations as *interstructuralism*.

7.3 Beyond reductionism and pluralism: Interstructuralism

The nature of the structural continuity between classical and quantum mechanics is not exhausted by the claim that the equations of the two theories are related by a limit (singular or otherwise); nor is it simply a matching of empirical predictions in some overlapping domain. When it comes to analyzing the continuity between these two theories, the sort of structures one is interested in are specifically *dynamical structures*. As we saw in some detail in Chapter 5, classical dynamical structures (such as periodic orbits and their stabilities) are not only manifesting themselves theoretically via a number of different correspondence principles, but are also manifesting themselves experimentally in the quantum phenomena. Once again, it is not that the quantum dynamics is mimicking the classical behavior in some limit, but rather that the classical dynamics is structuring the quantum phenomena in a number of surprising ways.

One of the lessons of Gutzwiller's trace formula is that classical dynamical structures can be used to construct (or model) quantum dynamical structures. In this case, it is not just a matter of two equations looking "similar" or being "partially isomorphic" to each other; rather, it is that the dynamical structures of one theory (e.g., classical periodic orbits) can be used to build up the dynamical structures of another theory (e.g., quantum density of states) in far more subtle and complex ways. It is precisely these sort of thoroughgoing structural correspondences that should be of interest to the scientific structuralist, and that are not captured by Nickles' reductionism$_2$ relation. An adequate account of the relation between these two theories should be able to incorporate the richer spectrum of correspondence relations revealed by semiclassical research.[26]

Drawing on Heinz Post's (1971) generalized correspondence principle, Simon Saunders rightly argues that much of the structural continuity one finds between classical and quantum mechanics can be understood on the basis of heuristics.[27] Saunders describes the role of heuristics in quantum mechanics as follows: "Innovation proceeds by isolation and independent development of structural features of extant theory. Once entrenched, such heuristics (or canonical forms) are preserved in subsequent developments, and previous theory reformulated in their terms" (Saunders 1993, p. 308). Saunders goes on to note many such examples of structural correspondences and heuristics in the development of quantum theory, including Dirac's correspondence principle. As we saw in the chapters on Heisenberg, Dirac, and Bohr, whether it was intentional or not, the founders of quantum theory used the dynamical structure of classical mechanics as the scaffolding on which quantum mechanics was built. Even Heisenberg – despite his rhetoric of making a clean break with classical mechanics and building the new quantum theory on observables only – makes use of the close correspondences between these theories. This does not, of course, mean that classical mechanics is preserved intact in quantum theory, only that it should not be surprising that significant remnants of these classical dynamical structures should still be found.

Not only were these correspondences critical to the development of quantum theory, but as we saw in Chapter 5, they are continuing to play a role in modern semiclassical research. Understanding the heuristic ground of these correspondences

[26] Further examples of these more complex correspondence relations can be found in Dirac ([1933] 2005), as we saw in Chapter 3.

[27] Recall that Post's generalized correspondence principle states that any new theory should be able to account for the successes of its predecessor by degenerating into it where the predecessor theory has been well-confirmed (Post 1971, p. 228). For a helpful disussion and history of the generalized correspondence principle, see Radder (1991).

can also help us recognize that intertheory relations are not static, but rather are evolving, and are continuing to be developed and extended in new ways. The recent developments in semiclassical mechanics, such as periodic orbit theory and closed orbit theory, have uncovered – and are continuing to uncover – a growing number of subtle and intricate correspondence relations between classical and quantum mechanics. As we saw in Chapter 3, Dirac repeatedly emphasized this evolving nature of intertheory relations, both in his account of classical and quantum mechanics as "open" theories and in his methodology of analogy extension. Indeed it is this emphasis on the dynamic nature of intertheory relations that sets Dirac's interstructuralism apart from traditional accounts of reductionism. Whether one theory logically entails another, or whether the equations of one theory are a limit of the equations of another theory, is something that may be difficult to discover; but when it is discovered, it is by and large decided once and for all by a simple yes or no. For the interstructuralist, by contrast, the structural correspondences between theories are open ended and continue to change with new developments and discoveries in mathematics and physics.

Interstructuralism is an approach to intertheory relations that emphasizes the importance of structural continuities and correspondences in giving an adequate account of the relation between two theories. It recognizes a richer diversity of correspondence relations than does any form of reductionism or pluralism. As we have seen, this richer set of correspondence relations is important not only theoretically, but also empirically for understanding the results of recent experiments in semiclassical research.

Applied to the case of classical and quantum mechanics, interstructuralism says that the relation between these theories consists in the fact that they to a large extent share the same dynamical structures, where that continuity of structure can be characterized by a number of different correspondence relations or correspondence principles. To say that an interstructuralist relation holds between two theories means that there is a close formal analogy between the two theories, which can provide the basis for a heuristic methodology of the sort described by Dirac.

Given the emphasis that I have placed here on dynamical structures it is not immediately clear how widely the interstructuralist approach to intertheory relations can be applied in describing other theory pairs. I suspect that it can be applied to closely related fields, such as chemistry, and perhaps be extended to other fields that possess something like a semiclassical mechanics, such as gravity, electromagnetism, and acoustics. Regrettably an exploration of this question lies outside the scope of this book. It would be surprising, however, if the same account of intertheory relations were to apply to all pairs of scientific theories. Which approach to intertheory relations is most appropriate depends on the concrete details of the theories being related. My approach here has been to try to argue that there are more ways of thinking about intertheory relations, not fewer.

Interstructuralism can be understood as a middle path between reductionism and theoretical pluralism.[28] From theoretical pluralism it takes the insight that predecessor or higher-level theories such as classical mechanics are still playing an important *theoretical* role in scientific research; that is, quantum mechanics – without classical mechanics – gives us an incomplete picture of our world. From reductionism, however, it takes the lesson that we cannot rest content with the view that each of these theories describes its own distinct domain of phenomena. We do not live in a dappled world, and we stand to miss out on many important scientific discoveries and insights if we do not try to bring our various theoretical descriptions of the world closer together.

7.4 Conclusion

Contemporary philosophy of science has for too long been dominated by what we might call the Heisenberg–Kuhn tradition, which views theories such as quantum mechanics as having made a radical break with their predecessors. This tradition takes the concepts of classical and quantum mechanics to be incommensurable with one another, and views quantum mechanics as an axiomatic system that is complete in itself. There is, however, an alternative tradition that might be called the Bohr–Dirac tradition, whose relevance for contemporary philosophy of science I have tried to revive here. This latter tradition sees quantum theory as a generalization of classical mechanics, though one that fully recognizes the contrast between these theories. It is thus an approach to the unity of science that, unlike other unities, still has room for differences. On this view, the concepts of classical and quantum mechanics are not incommensurable, but rather closely analogous; and for Bohr at least, quantum mechanics even requires the concepts of classical mechanics to be a complete theory. The Bohr–Dirac tradition emphasizes the many close correspondences between these theories and seeks to extend them in new ways.

Which of these two traditions one adopts is not just important for philosophy; they also each lay out a very different research program for physics. I have argued that the Bohr–Dirac tradition – and especially Dirac's particular expression of it – best describes the spirit underlying modern semiclassical research. Unlike Heisenberg and Pauli who thought new insights into physics would come only through working with our most fundamental physical theory, semiclassical theorists – along with Dirac – think that classical mechanics still has something more to teach us. More

[28] As such, it has some affinities with approaches to intertheory relations in the philosophy of biology such as Sandra Mitchell's (2003) "integrative pluralism."

specifically, one of the most important insights of semiclassical mechanics is that many new discoveries about quantum mechanics can be made by exploring its relation to classical mechanics. These are discoveries that it is unlikely one would have stumbled on just by considering quantum mechanics alone.[29] As I argued in Chapter 6, classical mechanics can often help us understand the structure of the quantum solutions in a way that the fully quantum approaches do not.

The semiclassical research described in Chapter 5 poses a number of interesting challenges for contemporary philosophy of science. I have considered only two here. The first is the challenge that semiclassical explanations pose for traditional philosophical accounts of scientific explanation. Specifically, the semiclassical appeals to classical structures in explaining quantum phenomena do not seem to fit well with either the deductive–nomological or causal accounts of explanation. I argued that a new account of scientific explanation is needed, which I called model explanations. The need for this new account of scientific explanation is by no means confined to semiclassical mechanics. Indeed there has been a growing recognition in the philosophy of science literature that scientific practice is to a large extent model based. If scientists are, as a matter of fact, using models to explain phenomena in nature, then we, as philosophers of science, should find a way of making sense of how that is possible. Although the task of developing a new account of scientific explanation, and responding to the long line of counterexamples that is sure to follow, is worthy of its own book(s), I outlined in Chapter 6 what I think is a promising line of approach to this challenge.

The second challenge raised by semiclassical research is to the adequacy of our current philosophical frameworks for thinking about intertheory relations. In particular, I argued that neither the traditional forms of reductionism nor the forms of theoretical pluralism adequately capture the relationship between classical and quantum mechanics. In searching for a more adequate approach to intertheory relations I turned to the philosophical views of Heisenberg, Dirac, and Bohr. This historical review revealed a number of surprising misconceptions about their views. Heisenberg is not a positivist, but rather a pluralistic realist; Bohr's correspondence principle is not actually about the asymptotic agreement of quantum and classical predictions; and Dirac does in fact have a philosophy of science. Although a proper understanding of these historical views is important in its own right, their views were also shown to parallel many of our contemporary debates in the philosophy of science – and perhaps more importantly, show how these debates might be moved forward.

[29] A very simple example of this is various quantum tunneling phenomena, which are defined precisely in relation to classical expectations. Many more examples can be found in the quantum chaos literature. See Casati and Chirikov (1995) for a review.

One topic I have not adequately addressed in this book is the issue of scientific realism. I have deliberately avoided talking about this topic, in part because the views in this debate tend to be so polarized that, once broached, they often crowd out other important issues in the philosophy of science. This was seen to be the case in the discussion of structural realism, for example, which, in its preoccupation with the realism question, has largely neglected the more fundamental continuity thesis on which it depends. The issue of scientific realism cannot be entirely avoided, however. As the preceding chapters reveal, the issues of intertheory relations, explanation, and realism are closely entwined.

The interstructuralist view outlined in this chapter can perhaps ameliorate what was likely for many a troubling claim in Chapter 6 that fictions can explain. If there is a continuity of structure between classical and quantum mechanics, and that structure can be adequately modeled by classical dynamical structures, then it is perhaps not so surprising that these classical structures – or rather what they represent – can successfully explain. Although the naive realist interpretation of these classical structures is blocked, there is room for a more sophisticated realism about mathematical structures. Whether this continuity of dynamical structure that interstructuralism emphasizes is indicative of its representing a real structure out there in the world as the realist contends, or whether it is merely an artifact of the heuristic strategies employed in the development of these theories, as the antirealist contends, is not clear. I do not see either side in this debate coming closer to convincing the other. What I think interstructuralism does clearly show, however, is that, in response to the question posed in the title of Chapter 1, imperialism and isolationism are not our only options.

References

Adler, S. (2003), Why decoherence has not solved the measurement problem: A response to P. W. Anderson, *Studies in History and Philosophy of Modern Physics* **34**B: 135–42.

Bacciagaluppi, G. (2004), The role of decoherence in quantum mechanics, in E.N. Zalta (ed.) *The Stanford Encyclopedia of Philosophy*. http://plato.stanford.edu/archives/sum 2005/entries/qm-decoherence/.

Baker, A. (2005), Are there genuine mathematical explanations of physical phenomena?, *Mind* **114**: 223–38.

Ballentine, L. (1970), The statistical interpretation of quantum mechanics, *Reviews of Modern Physics* **42**: 358–381. Reprinted in L. Ballentine (1988), *Foundations of Quantum Mechanics since the Bell Inequalities, Selected Reprints*. College Park: American Association of Physics Teachers, pp. 9–32.

Ballentine, L. (2004), Quantum-to-classical limit in a Hamiltonian system, *Physical Review A* **70**: 032111–1 to 032111–7.

Ballentine, L., Y. Yang, and J.P. Zibin (1994), Inadequacy of Ehrenfest's theorem to characterize the classical regime, *Physical Review A*: **50**: 2854–9.

Baranger, H., R. Jalabert, and D. Stone (1993), Quantum-chaotic scattering effects in semiconductor microstructures, *Chaos*, **3**: 665–82.

Batterman, R. (1991), Chaos, quantization, and the correspondence principle, *Synthese*, **89**: 189–227.

Batterman, R. (1992), Explanatory instability, *Noûs* **26**: 325–48.

Batterman, R. (1993), Quantum chaos and semiclassical mechanics, *PSA 1992, Proceedings of the Biennial Meeting of the Philosophy of Science Association*, Vol. Two: Symposia and Invited Papers, 50–65.

Batterman, R. (1995), Theories between theories: asymptotic limiting intertheoretic relations, *Synthese* **103**: 171–201.

Batterman, R. (2002), *The Devil in the Details: Asymptotic Reasoning in Explanation, Reduction and Emergence*. Oxford: Oxford University Press.

Batterman, R. (2005), Response to Belot's "Whose devil? Which details?", *Philosophy of Science*, **72**: 154–63.

Beller, M. (1996), The rhetoric of antirealism and the Copenhagen spirit, *Philosophy of Science* **63**: 183–204.

Beller, M. (1999), *Quantum Dialogue: The Making of a Revolution*. Chicago: University of Chicago Press.

Belot, G. (2000), Chaos and fundamentalism, *Philosophy of Science* **67** (Proceedings): S454–S465.

Belot, G. (2005), Whose devil? Which details?, *Philosophy of Science*, **72**: 128–53.

Berry, M. V. (1977), Regular and irregular semiclassical wavefunctions, *Journal of Physics A* **10** (12): 2083–91.

Berry, M. V. (1982), Semiclassical weak reflections above analytic and nonanalytic potential barriers, *Journal of Physics A* **15**: 3693–704.

Berry, M. V. (1983), Semiclassical mechanics of regular and irregular motion, in G. Iooss, R. H. G. Helleman, and R. Stora, (eds.), *Les Houches Lecture Series Session XXXVI*, Amsterdam: North-Holland, 171–271.

Berry, M. V. (1987), Quantum chaology (The Bakerian Lecture), *Proceedings of the Royal Society A* 413: 183–98.

Berry, M. V. (1988), Random renormalization in the semiclassical long-time limit of a precessing spin, *Physica D* **33**: 26–33.

Berry, M. V. (1989), Quantum chaology, not quantum chaos, *Physica Scripta* **40**: 335–6.

Berry, M. V. (1991), Some quantum-to-classical asymptotics, in M. J. Giannoni, A. Voros, and J. Zinn-Justin (eds.) *Chaos and Quantum Physics*. Amsterdam: North-Holland, 250–303.

Berry, M. V. (1994), Asymptotics, singularities and the reduction of theories, in D. Prawitz, B. Skyrms, and D. Westerstahl (eds.) *Logic, Methodology and Philosophy of Science IX*. Amsterdam: Elsevier.

Berry, M. V. (2001), Chaos and the semiclassical limit of quantum mechanics (Is the moon there when somebody looks?), in R. J. Russell, P. Clayton, K. Wegter-McNelly, and J. Polkinghorne (eds.) *Quantum Mechanics: Scientific Perspectives on Divine Action*. Vatican Observatory: CTNS Publications.

Berry, M. V. (2002), Singular limits, *Physics Today*, **55**(5): 10–11.

Berry, M. V. and K. Mount (1972), Semiclassical approximations in wave mechanics, *Reports on Progress in Physics*, **35**: 315–97.

Bohr, N. (1913), On the constitution of atoms and molecules, *Philosophical Magazine* **26**: 1–25, 476–502, 857–75.

Bohr, N. (1918), The quantum theory of line-spectra, *D. Kgl. Danske Vidensk. Selsk. Skrifter, naturvidensk. og mathem*. Afd. 8. Reikke, IV.1, reprinted in Bohr (1976), pp. 67–102.

Bohr, N. (1920), On the series spectra of the elements, Lecture before the German Physical Society in Berlin (27 April 1920), translated by A. D. Udden, in Bohr (1976), 241–82.

Bohr, N. (1921a), Foreword to German translation of early papers on atomic structure, in Bohr (1976), 325–337; translation of "Geleitwort" in H. Stintzing (ed.) *Abhandlungen űber Atombau aus den Jahren 1913–1916*, Braunschweig: F. Vieweg & Sohn (1921).

Bohr, N. (1921b), Constitution of atoms, in Bohr (1977), pp. 99–174.

Bohr, N. (1922), Seven lectures on the theory of atomic structure, in Bohr (1977), pp. 341–419.

Bohr, N. (1924), On the application of the quantum theory to atomic structure, *Proceedings of the Cambridge Philosophical Society (suppl.)*. Cambridge: Cambridge University Press, pp. 1–42. Reprinted in Bohr (1976), pp. 457–99.

Bohr, N. (1925), Atomic theory and mechanics, *Nature (suppl.)* **116**, 845–852. Reprinted in Bohr (1984), pp. 273–80.

Bohr, N. (1928), The quantum postulate and the recent development of atomic theory, *Nature (suppl.)* **121**: 580–590. Reprinted Bohr (1985), pp. 148–58.

Bohr, N. ([1929] 1934) Introductory survey, in *Atomic Theory and the Description of Nature*, (Cambridge University Press, Cambridge, [1929] 1934), pp. 1–24. Reprinted in Bohr (1985), pp. 279–302.

Bohr, N. (1931), Maxwell and modern theoretical physics, *Nature* **128** (3234): 691–692. Reprinted in Bohr (1985), pp. 359–60.

Bohr, N. (1935), Can quantum-mechanical description of physical reality be considered complete?, *Physical Review* **48**: 696–702. Reprinted in Bohr (1996), pp. 292–8.

Bohr, N. (1948), On the notions of causality and complementarity, *Dialectica* **2**, 312–319. Reprinted in Bohr (1996), pp. 330–7.

Bohr, N, (1949), Discussion with Einstein on epistemological problems in atomic physics, in P.A. Schilpp (ed.) *Albert Einstein: Philosopher-Scientist*, The Library of Living Philosophers, Vol. VII. La Salle, IL: Open Court, pp. 201–41. Reprinted in Bohr (1996), pp. 341–81.

Bohr, N. (1958b), Quantum physics and philosophy: causality and complementarity, in R. Klibansky, (ed.) *Philosophy in the Mid-Century: A Survey*. Firenze: La Nuova Italia Editrice. Reprinted in Bohr (1996), pp. 388–94.

Bohr N. (1976), *Niels Bohr Collected Works, Vol. 3: The Correspondence Principle (1918–1923)*, J.R. Nielsen (ed.). Amsterdam: North-Holland Publishing.

Bohr N. (1977), *Niels Bohr Collected Works, Vol. 4: The Periodic System (1920–1923)*, J.R. Nielsen (ed.). Amsterdam: North-Holland Publishing.

Bohr, N. (1984), *Niels Bohr Collected Works, Vol. 5: The Emergence of Quantum Mechanics (Mainly 1924–1926)*, K. Stolzenburg (ed.). Amsterdam: North-Holland Publishing.

Bohr N. (1985), *Niels Bohr Collected Works, Vol. 6: Foundations of Quantum Physics I (1926–1932)*, J. Kalckar (ed.). Amsterdam: North-Holland Publishing.

Bohr, N. (1996), *Niels Bohr Collected Works, Vol. 7: Foundations of Quantum Physics II (1933–1958)*, J. Kalckar (ed.). Amsterdam: North-Holland Publishing.

Bokulich, A. (2001) Philosophical perspectives on quantum chaos: Models and interpretations. Ph.D. Dissertation, University of Notre Dame.

Bokulich, A. (2004), Open or closed? Dirac, Heisenberg, and the relation between classical and quantum mechanics, *Studies in History and Philosophy of Modern Physics* **35**: 377–96.

Bokulich, A. (2006), Heisenberg meets Kuhn: Closed theories and paradigms, *Philosophy of Science* **73**: 90–107.

Bokulich, A. (2008), Can classical structures explain quantum phenomena?, *British Journal for the Philosophy of Science* **59** (2): 217–35.

Bokulich, A. (2008), Paul Dirac and the Einstein-Bohr debate, *Perspectives on Science* **16**(1): 103–14.

Bokulich, P. and A. Bokulich (2005), Niels Bohr's generalization of classical mechanics, *Foundations of Physics* **35**: 347–71.

Born, M. (1926), *Problems of Atomic Dynamics*. Cambridge, MA: MIT Press.

Born, M. and W. Heisenberg ([1927] 1928), La mécanique des quanta, in Institut International de Physique Solvay (ed.) *Électrons et Photons: Rapports et Discussions du Cinquième Conseil de Physique Tenu à Bruxelles du 24 au 29 Octobre 1927 sous les Auspices de l'Institut International de Physique Solvay*. Paris: Gauthier-Villars, 143–84. Reprinted in Heisenberg (1984), 58–96.

Brillouin, L. (1926), La mécanique ondulatoire de Schrödinger; Une méthod général de résolution par approximations successives, *Comptes Rendus de Séances de l'Academie des Sciences*, **183**: 24–6.

Bueno, O. (1997), Empirical adequacy: A partial structures approach, *Studies in History and Philosophy of Science* **28**: 585–610.

Bueno, O. (1999), What is structural empiricism? Scientific change in an empiricist setting, *Erkenntnis* **50**: 59–85.

Bueno, O. (2000), Empiricism, scientific change and mathematical change, *Studies in History and Philosophy of Science* **31**: 269–96.

Bueno, O., S. French, and J. Ladyman (2002), On representing the relationship between the mathematical and the empirical, *Philosophy of Science* **69**: 452–73.

Cartwright, N. (1995), The metaphysics of the disunified world, in D. Hull, M. Forbes, and R. Burian (eds.) *PSA 1994*, vol. 2. East Lansing, MI: Philosophy of Science Association, pp. 357–64.

Cartwright, N. (1999), *The Dappled World: A Study of the Boundaries of Science*. Cambridge: Cambridge University Press.

Casati, G. and B. Chirikov (1995), The legacy of chaos in quantum mechanics, in G. Casati and B. Chirikov (eds.) *Quantum Chaos: Between Order and Disorder*. Cambridge: Cambridge University Press, pp. 3–53.

Cassidy, D. (1992), *Uncertainty: The Life and Science of Werner Heisenberg*. New York: Freeman Press.

Chevalley, C. (1988), Physical reality and closed theories in Werner Heisenberg's early papers, in D. Batens and J.P. van Bendegem (eds.) *Theory and Experiment: Recent Insights and New Perspectives on their Relation*. Dordrecht, Reidel, 159–76.

Clifton, R. (1998) Scientific explanation in quantum theory, Unpublished manuscript on PhilSci Archive: http://philsci-archive.pitt.edu/archive/00000091/.

Craver, C. (2006), When mechanistic models explain, *Synthese* **153**: 355–76.

Crowe, M.J. (1990), *Theories of the World from Antiquity to the Copernican Revolution*. New York: Dover.

Cvitanović, P., R. Artuso, R. Mainieri, G. Tanner, G. Vattay, N. Whelan, and A. Wirzba (2005), *Chaos: Classical and Quantum*. http://ChaosBook.org/version11 (Niels Bohr Institute, Copenhagen).

da Costa, N. and S. French (2000), Models, theories, and structures: Thirty years on, *Philosophy of Science* **67**: S116–S127.

da Costa N. and S. French (2003), *Science as Partial Truth: A Unitary Approach to Models and Scientific Reasoning*. Oxford: Oxford University Press.

Darrigol, O. (1992), *From c-Numbers to q-Numbers: The Classical Analogy in the History of Quantum Theory*. Berkeley: University of California Press.

Darrigol, O. (1997), Classical concepts in Bohr's atomic theory (1913–1925), *Physis: Rivista Internazionale di Storia della Scienza* **34**: 545–67.

Delos, J.B., S.K. Knudson and D.W. Noid (1983), Highly excited states of a hydrogen atom in a strong magnetic field, *Physical Review A* **28**: 7–21.

Delos, J.B., and M.-L. Du (1988), Correspondence principles: the relationship between classical trajectories and quantum spectra, *IEEE Journal of Quantum Electronics* **24**: 1445–52.

Dirac, P.A.M. (1925), The fundamental equations of quantum mechanics, *Proceedings of the Royal Society of London, Series A*, **109**: 642–53.

Dirac, P.A.M. (1926a), Quantum mechanics and preliminary investigation of the hydrogen atom, *Proceedings of the Royal Society of London, Series A*, **110**: 561–79.

Dirac, P.A.M. (1926b), The elimination of the nodes in quantum mechanics, *Proceedings of the Royal Society of London, Series A*, **110**: 281–305.

Dirac, P.A.M. (1932), Relativistic quantum mechanics, *Proceedings of the Royal Society of London. Series A*, **136**: 453–64. Reprinted in Dirac (1995), pp. 621–34.

Dirac, P.A.M. ([1933] 2005), The Lagrangian in quantum mechanics, *Physikalische Zeitschrift der Sowjetunion*, **3**.1 (1933): 64–72. Reprinted in L. M. Brown (ed.) (2005), *Feynman's Thesis: A New Approach to Quantum Theory*. London: World Scientific, pp. 113–21.

Dirac, P.A.M. (1938), Classical theory of radiating electrons, *Proceedings of the Royal Society of London, Series A*, **167**, 148–69.

Dirac, P.A.M. (1939), The relation between mathematics and physics, *Proceedings of the Royal Society (Edinburgh)*, **59**: 122–9. Reprinted in Dirac (1995), pp. 905–14.

Dirac, P.A.M. (1945), On the analogy between classical and quantum mechanics, *Reviews of Modern Physics*, **17**, 195–99.

Dirac, P.A.M. ([1948] 1995), "Quelques développements sur la théorie atomique" Conférence faites an Palais de la Découverte, 6 December 1945. Paris: Université de Paris, 1948. Translated and reprinted in Dirac (1995), 1246–67.

Dirac, P.A.M. (1951a), The relation of classical to quantum mechanics, *Proceedings of the Second Mathematical Congress, Vancouver, 1949*. Toronto: University of Toronto Press, pp. 10–31.

Dirac, P.A.M. (1951b), Is there an aether?, *Nature*, **168**, 906–7.

Dirac, P.A.M. (1952), Is there an aether?, *Nature*, **169**, 702.

Dirac, P.A.M. (1953), The Lorentz transformation and absolute time, *Physica*, **19**, 888–96.

Dirac, P.A.M. (1958), *The Principles of Quantum Mechanics* (4th edn.). Oxford: Clarendon Press.

Dirac, P. (1962, April 1st), Oral history interview of P.A.M. Dirac by Thomas Kuhn. Archive for the History of Quantum Physics, deposit at Harvard University, Cambridge, MA.

Dirac, P.A.M. (1963, May 14th), Oral history interview of P.A.M. Dirac by Thomas Kuhn. Archive for the History of Quantum Physics, deposit at Harvard University, Cambridge, MA.

Dirac, P.A.M. (1963), The evolution of the physicist's picture of nature, *Scientific American*, **208**(5), 45–53.

Dirac, P.A.M. (1970), Can equations of motion be used in high-energy physics?, *Physics Today*, April, 29–31.

Dirac, P.A.M. (1971), *The Development of Quantum Theory: J. Robert Oppenheimer Memorial Prize Acceptance Speech*. New York: Gordon and Breach Science Publishers.

Dirac, P.A.M. (1974), Einstein and Bohr: The great controversy, Unpublished lecture. Paul A. M. Dirac Collection, Series 2, Box 29, Folder 4, Florida State University, Tallahassee, Florida, USA.

Dirac, P.A.M. (1977), Recollections of an exciting era, in C. Weiner (ed.) *History of Twentieth Century Physics, Proceedings of the International School of Physics "Enrico Fermi" 1972*. New York: Academic Press, pp. 109–46.

Dirac, P.A.M. ([1979] 1982), The early years of relativity, in G. Holton and Y. Elkana (eds.) *Albert Einstein: Historical and Cultural Perspectives, Jerusalem Einstein Centennial Symposium 14–23 March 1979* (pp. 79–90). Princeton, NJ: Princeton University Press.

Dirac, P.A.M. (1987), The inadequacies of quantum field theory, in B. Kursunoglu and E. Wigner (eds.) *Reminiscences About a Great Physicist: P.A.M. Dirac*. Cambridge: Cambridge University Press, pp. 194–8.

Dirac, P.A.M. (1995), *The Collected Works of P. A. M. Dirac: 1924–1948*, edited by R.H. Dalitz. Cambridge: Cambridge University Press.

Du, M.-L., and J.B. Delos (1988), Effect of closed classical orbits on quantum spectra: Ionization of atoms in a magnetic field. I. Physical picture and calculations, *Physical Review A* **38**: 1896–912.

Dupré, J. (1995), Against scientific imperialism, in D. Hull, M. Forbes, and R. Burian (eds.), *PSA 1994*, vol. 2, East Lansing, MI: Philosophy of Science Association, 374–81.

Dupré, J. (1996a), *The Disorder of Things: Metaphysical Foundations of the Disunity of Science*. Cambridge, MA: Harvard University Press.

Dupré, J. (1996b), Metaphysical disorder and scientific disunity, in P. Galison and D. Stump (eds.), *The Disunity of Science*. Stanford: Stanford University Press, 101–17.

Einstein, A. (1917), Zum Quantensatz von Sommerfeld und Epstein, *Verhandlungen der Deutschen Physikalischen Gesellschaft*, **19**: 82–92. Translated into English and reprinted in Einstein (1997), pp. 434–43.

Einstein, A. (1926), [Letter to Max Born on December 4, 1926]. Translated and reprinted in *The Born-Einstein Letters: Correspondence between Albert Einstein and Max and Hedwig Born from 1916 to 1955 with Commentaries by Max Born*. New York: Walker and Co., 90–1.

Einstein, A. ([1927] 1928), Comments in "Discussion générale des idées nouvelles émises". *Électrons et photons: rapports et discussions du cinquième conseil de physique, tenu à Bruxelles du 24 and 29 Octobre 1927, sous les auspices de l'Institut International de Physique Solvay.* Paris: Gauthier-Villars et Cie. Reprinted in J. Kalckar (ed.), *Niels Bohr Collected Works, Volume 6: Foundations of Quantum Physics I 1926–1932).* Amsterdam: North-Holland, 1985, 101–3.

Einstein, A. ([1933] 1934), On the method of theoretical physics, The Herbert Spencer Lecture, Delivered at Oxford, June 10, 1933. Reproduced in *Philosophy of Science* **1**: 163–9.

Einstein, A. ([1949] 1970), Remarks to the essays appearing in this volume, in P. Schilpp (ed), Albert Einstein: Philosopher-Scientist. La Salle: Open Court. pp. 663–88.

Einstein, A. (1997), *The Collected Papers of Albert Einstein*, vol. 6. A. Engel (trans.), Princeton: Princeton University Press.

Elgin, M. and E. Sober (2002), Cartwright on explanation and idealization, *Erkenntnis* **57**: 441–50.

Emerson, J. (2001), Chaos and quantum-classical correspondence for coupled spins, Ph.D. Dissertation, Simon Fraser University.

Epstein, P. (1916), Zur Quantentheorie, *Annalen der Physik*, **51**: 168–88.

Ezra, G.S. (1998), Classical-quantum correspondence and the analysis of highly excited states: Periodic orbits, rational tori, and beyond, in W. Hase (ed.) *Advances in Classical Trajectory Methods*, Vol. 3. Greenwich, CT: JAI Press, pp. 35–72.

Ezra, G.S., K. Richter, G. Tanner, and D. Wintgen (1991), Semiclassical cycle expansion for the helium atom, *Journal of Physics B*, **24**: L413–L420.

Faye, J. and H. Folse (eds.) (1994), *Niels Bohr and Contemporary Philosophy*, Boston Studies in the Philosophy of Science, Vol. 153. Dordrecht: Kluwer Academic.

Fedak, W.A., and J.J. Prentis (2002), Quantum jumps and classical harmonics, *American Journal of Physics* **70**: 332–44.

Feyerabend, P. (1962), Explanation, reduction, and empiricism, in H. Feigl and G. Maxwell (eds.) *Minnesota Studies in the Philosophy of Science*. Minneapolis: University of Minnesota Press, pp. 28–97.

Feyerabend, P. ([1965] 1983), Problems of empiricism, in R. Colodny (ed.) *Beyond the Edge of Certainty: Essays in Contemporary Science and Philosophy*. Lanham, MD: University Press of America, pp. 145–260.

Feynman, R. (1948), Space-time approach to non-relativistic quantum mechanics, *Reviews of Modern Physics* **10**: 367–87.

Feynman, R. (1965), The development of the space-time view of quantum electrodynamics, Nobel Lecture. http://nobelprize.org/nobel_prizes/physics/laureates/1965/feynman-lecture.html

Fine, A. (1996), *The Shaky Game: Einstein, Realism and The Quantum Theory*, 2nd edition, Chicago: University of Chicago Press.

Fodor, J. (1974), Special sciences (or: The disunity of science as a working hypothesis), *Synthese* **28**: 97–115.

Fodor, J. (1997), Special sciences: Still autonomous after all these years, *Noûs* **31**, Supplement: *Philosophical Perspectives, 11, Mind, Causation, and World*, 149–63.

Ford, J. and G. Mantica (1992), Does quantum mechanics obey the correspondence principle? Is it complete?, *American Journal of Physics* **60**: 1086–98.

Frappier, M. (2004), *Heisenberg's Notion of Interpretation*. Ph.D. Dissertation. London, Ontario: The University of Western Ontario.

French, S. (1997), Partiality, pursuit and practice, in M.L. dalla Chiara, K. Doets, D. Mundici, and J. van Benthem (eds.) *Logic and Scientific Methods*, Volume One of the Tenth International Congress of Logic, Methodology and Philosophy of Science, Florence, August 1995. Dordrecht: Kluwer Academic, pp. 35–52.

French, S. (2000), The reasonable effectiveness of mathematics: Partial structures and the application of group theory to physics, *Synthese* **125**: 103–20.

French, S. and J. Ladyman (1999), Reinflating the model-theoretic approach, *International Studies in the Philosophy of Science* **13**: 99–117.

French, S. and J. Ladyman (2003), Remodelling structural realism: Quantum physics and the metaphysics of structure, *Synthese* **136**: 31–56.

Friedman, M. (1974), Explanation and scientific understanding, *Journal of Philosophy* **71**: 5–19.

Frisch, M. (2002), Non-locality in classical electrodynamics, *British Journal for the Philosophy of Science* **53**: 1–19.

Frisch, M. (2005), *Inconsistency, Asymmetry, and Non-Locality: A Philosophical Investigation of Classical Electrodynamics*. Oxford: Oxford University Press.

Galison, P. (1997), *Image and Logic: A Material Culture of Microphysics*. Chicago: University of Chicago Press.

Garton, W. and F. Tomkins (1969), Diamagnetic Zeeman effect and magnetic configuration mixing in long spectral series of Ba I, *The Astrophysical Journal* **158**: 839–45.

Granger, B.E. (2001), *Quantum and Semiclassical Scattering Matrix Theory for Atomic Photoabsorption in External Fields*, Ph.D. Thesis, University of Colorado.

Gutzwiller, M. (1971), Periodic orbits and classical quantization conditions, *Journal of Mathematical Physics* **12**: 343–58.

Gutzwiller, M. (1990), *Chaos in Classical and Quantum Mechanics*. New York: Springer-Verlag.

Habib, S., K. Shizume, and W. Zurek (1998), Decoherence, chaos, and the correspondence principle, *Physical Review Letters* **80**: 4361–5.

Haggerty, M.R., N. Spellmeyer, D. Kleppner, and J.B. Delos (1998), Extracting classical trajectories from atomic spectra, *Physical Review Letters* **81**: 1592–5.

Hassoun, G. and D. Kobe (1989), Synthesis of the Planck and Bohr formulations of the correspondence principle, *American Journal of Physics* **57**(7): 658–62.

Heisenberg, W. ([1925] 1967), Quantum-theoretical re-interpretation of kinematic and mechanical relations, in B. van der Waerden (ed.) *Sources of Quantum Mechanics*. New York: Dover Publications, 261–276. Translation of "Über quantentheorische Umdeutung kinematischer und mechanischer Beziehungen", *Zeitschrift für Physik* **33**: 879–93.

Heisenberg, W. ([1927] 1983), The physical content of quantum kinematics and mechanics, translated and reprinted in J.A. Wheeler and W.H. Zurek (eds.) *Quantum Theory and Measurement*. Princeton, NJ: Princeton University Press, pp. 62–84.
Heisenberg, W. (1930), *The Physical Principles of the Quantum Theory*. Translated by C. Eckart and F.C. Hoyt. Chicago: University of Chicago Press.
Heisenberg, W. ([1934] 1979), Recent changes in the foundations of exact science, in F.C. Hayes (trans.) *Philosophical Problems of Quantum Physics*. Woodbridge, CT: Ox Bow Press, pp. 11–26.
Heisenberg, W. ([1935] 1979), Questions of principle in modern physics, Lecture delivered at Vienna University on November 27, 1935. Published in W. Heisenberg, *Philosophical Problems of Quantum Physics*. Woodbridge, CT: Ox Bow Press, 41–52.
Heisenberg, W. ([1938] 1994), The universal length appearing in the theory of elementary particles, in A.I. Miller (ed.) *Early Quantum Electrodynamics: A Sourcebook*. Cambridge, Cambridge University Press, pp. 244–53.
Heisenberg, W. ([1941a] 1979), The teachings of Goethe and Newton on colour in light of modern physics, in *Philosophical Problems of Quantum Physics*. Woodbridge, CT: Ox Bow Press, pp. 60–76.
Heisenberg, W. ([1941b] 1990), On the unity of the scientific outlook on nature, in *Philosophical Problems of Quantum Physics*. Woodbridge, CT: Ox Bow Press, pp. 77–94.
Heisenberg, W. ([1942] 1998), *Ordnung der Wirklichkeit*. Translated into French by Catherine Chevalley as *Philosophie: Le Manuscrit de 1942*. Paris: Éditions du Seuil.
Heisenberg, W. ([1948] 1974), The notion of a "closed theory" in modern science, in W. Heisenberg *Across The Frontiers*. New York: Harper & Row, Publishers, 39–46. Originally published as "Der Begriff 'Abgeschlossene Theorie' in der Modernen Naturwissenschaft", *Dialectica* **2**: 331–6.
Heisenberg, W. (1958a), *Physics and Philosophy: The Revolution in Modern Science*. New York: Harper & Row.
Heisenberg, W. ([1958b] 1974), Planck's discovery and the philosophical problems of atomic theory, in *Across The Frontiers*. New York: Harper & Row, pp. 8–29.
Heisenberg, W. (1961), Goethe's view of nature and the world, in A.B. Wachsmuth (ed.) *Goethe–New Series of the Goethe Society Yearbook*, Vol. 29. Weimar: Hermann Bohlaus Nachfolger, pp. 27–42. Reprinted in Heisenberg ([1970] 1990), pp. 121–41.
Heisenberg, W. (1963, February 15th), Oral history interview of Werner Heisenberg by Thomas Kuhn. Archive for the History of Quantum Physics, deposit at Harvard University, Cambridge, MA.
Heisenberg, W. (1963, February 27th), Oral history interview of Werner Heisenberg by Thomas Kuhn. Archive for the History of Quantum Physics, deposit at Harvard University, Cambridge, MA.
Heisenberg, W. (1963, February 28th), Oral history interview of Werner Heisenberg by Thomas Kuhn. Archive for the History of Quantum Physics, deposit at Harvard University, Cambridge, MA.
Heisenberg, W. (1963, July 5th), Oral history interview of Werner Heisenberg by Thomas Kuhn. Archive for the History of Quantum Physics, deposit at Harvard University, Cambridge, MA.
Heisenberg, W. (1966), Die Rolle der phänomenologischen Theorien im System der theoretischen Physik, in A. De-Shalit, H. Feshbach, and L. Van Hove (eds.) *Preludes in Theoretical Physics: In Honor of V. F. Weisskopf*. Amsterdam: North-Holland Publishing Co., pp. 166–169. Reprinted in Heisenberg (1984), 384–7.

Heisenberg, W. (1968), Theory, criticism and a philosophy, in *From a Life Of Physics: Evening Lectures at the International Centre for Theoretical Physics, Trieste, Italy.* Wien: IAEA, 31–46. Reprinted in Heisenberg (1984), 423–38.
Heisenberg, W. ([1969] 1990), Changes of thought pattern in the progress of science, in Heisenberg ([1970] 1990), pp. 154–65.
Heisenberg, W. ([1970] 1990), *Across the Frontiers*, translated from German by Peter Heath. Woodbridge, CT: Ox Bow Press.
Heisenberg, W. (1971), Atomic physics and pragmatism (1929), in *Physics and Beyond: Encounters and Conversations*. New York: Harper & Row, pp. 93–102.
Heisenberg, W. ([1972] 1983), The correctness-criteria for closed theories in physics, in W. Heisenberg *Encounters with Einstein: And Other Essays on People, Places, and Particles*. Princeton, NJ: Princeton University Press, pp. 123–9.
Heisenberg, W. ([1952] 1979), *Philosophical Problems of Quantum Physics*. Woodbridge, CT: Ox Bow Press.
Heisenberg, W. (1984), *Gesammelte Werke/Collected Works*, Volume C2, W. Blum, H.-P. Durr, and H. Rechenberg (eds.). Munchen: Piper.
Heller, E.J. (1984), Bound-state eigenfunctions of classically chaotic Hamiltonian systems: scars of periodic orbits, *Physical Review Letters*, **53**: 1515–18.
Heller, E.J. (1986), Qualitative properties of eigenfunctions of classically chaotic Hamiltonian systems in T.H. Seligman and H. Nishioka (eds.) *Quantum Chaos and Statistical Nuclear Physics*, New York: Springer-Verlag, pp. 162–81.
Heller, E. and S. Tomsovic (1993), Postmodern quantum mechanics, *Physics Today*, **46** (July): 38–46.
Hempel, C. (1965), *Aspects of Scientific Explanation and Other Essays in the Philosophy of Science*. New York: Free Press.
Hesse, M.B. (1966), *Models and Analogies in Science*. Notre Dame, IN: University of Notre Dame Press.
Hitchcock, C. and J. Woodward (2003), Explanatory generalizations, part II: Plumbing explanatory depth, *Noûs* **37**: 181–99.
Holle, A., J. Main, G. Wiebusch, H. Rottke, and K.H. Welge (1988), Quasi-Landau spectrum of the chaotic diamagnetic hydrogen atom, *Physical Review Letters* **61**: 161–4.
Hooker, C. (2004), Asymptotics, reduction and emergence, *British Journal for the Philosophy of Science*, **55**: 435–79.
Howard, D. (1990). "Nicht sein kann was nicht sein darf," or the prehistory of EPR, 1909–1935: Einstein's early worries about the quantum mechanics of composite systems. In A.I. Miller (ed.) *Sixty-Two Years of Uncertainty: Historical, Philosophical and Physical Inquiries into to the Foundations of Quantum Mechanics*. New York: Plenum, pp. 61–111.
Howard, D. (1994), What makes a classical concept classical? Towards a reconstruction of Niels Bohr's philosophy of physics, in J. Faye and H. Folse (eds.) *Niels Bohr and Contemporary Philosophy*, Boston Studies in the Philosophy of Science, Vol. **153**. Dordrecht: Kluwer Academic, pp. 201–29.
Howard, D. (2004), Who invented the Copenhagen Interpretation? A study in mythology, *Philosophy of Science* **71**: 669–82.
Hughes, R.I.G. (1989), Bell's Theorem, ideology, and structural explanation, *Philosophical Consequences of Quantum Theory: Reflections on Bell's Theorem*. Notre Dame, IN: University of Notre Dame Press, 195–207.
Hughes, R.I.G. (1993), Theoretical explanation, *Midwest Studies in Philosophy* Vol. **XVIII**: 132–53.

Hylleraas, E. (1963), Reminiscences from early quantum mechanics of two-electron atoms, *Reviews of Modern Physics*, **35** (3): 421–31.

Jalabert, R. (2000), The semiclassical tool in mesoscopic physics, in G. Casati, I. Guarneri, and U. Smilansky (eds.) *New Directions in Quantum Chaos*, Proceedings of the International School of Physics 'Enrico Fermi', Course CXLIII. Amsterdam: IOS Press, 145–222.

Jammer, M. (1966), *The Conceptual Development of Quantum Mechanics*. New York: McGraw Hill Book Co.

Jammer, M. (1974), *The Philosophy of Quantum Mechanics*. New York: Wiley.

Jensen, R.V. (1992), Quantum chaos, *Nature*, **355**: 311–18.

Kaplan, L. (1999), Scars in quantum chaotic wavefunctions, *Nonlinearity* **12**: R1–R40.

Keller, J. (1958), Corrected Bohr-Sommerfeld quantum conditions for nonseperable systems, *Annals of Physics*, **4**: 180–8.

Keller, J. (1985), Semiclassical mechanics, *SIAM Review*, **27** (4): 485–504.

Kellert, S., H. Longino and C.K. Waters (eds.) (2006), *Scientific Pluralism*, Minnesota Studies in the Philosophy of Science, XIX. Minneapolis: University of Minnesota Press.

Kemeny, J. and P. Oppenheim. (1956), On reduction, *Philosophical Studies* **7**: 6–19.

Kitcher, P. (1984), 1953 and all that. A tale of two sciences, *The Philosophical Review*, **93**: 335–73.

Klavetter, J. (1989a), Rotation of Hyperion. I–Observations, *Astronomical Journal* **97**: 570–9.

Klavetter, J. (1989b), Rotation of Hyperion. II–Dynamics, *Astronomical Journal* **98**: 1855–74.

Kleppner, D. (1991), Quantum chaos and the bow-stern enigma, *Physics Today*, August: 10–11.

Kleppner, D. and J.B. Delos (2001), Beyond quantum mechanics: insights from the work of Martin Gutzwiller, *Foundations of Physics* **31**: 593–612.

Koopman, B.O. (1931), Hamiltonian systems and transformations in Hilbert space, *Proceedings of the National Academy of Sciences* **18**: 315–18.

Kragh, H. (1990). *Dirac: A Scientific Biography*. Cambridge: Cambridge University Press.

Kramers, H. (1926), Wellenmechanik und halbzahlige Quantisierung, *Zeitschrift fur Physik*, **39**: 828–40.

Kuhn, T. ([1962] 1996), *The Structure of Scientific Revolutions*, 3rd edn. Chicago: University of Chicago Press.

Kuhn, T. (1963, February 27th), Oral history interview of Werner Heisenberg by Thomas Kuhn. Archive for the History of Quantum Physics, deposit at Harvard University, Cambridge, MA.

Kuhn, T. (1963, February 28th), Oral history interview of Werner Heisenberg by Thomas Kuhn. Archive for the History of Quantum Physics, deposit at Harvard University, Cambridge, MA.

Ladyman, J. (1998), What is structural realism?, *Studies in History and Philosophy of Science* **29**: 409–24.

Lakatos, I. (1970), Falsification and the Methodology of scientific Research Programmes, in I. Lakatos and A. Musgrave (eds.) *Criticism and the Growth of Knowledge*. Cambridge: Cambridge University Press, 91–196.

Landau, L.D. and E.M. Lifshitz (1981), *Mechanics*, 3rd edn. Oxford: Butterworth-Heinemann.

Landsman, N.P. (2007), Between classical and quantum, in J. Butterfield and J. Earman (eds.) *Philosophy of Physics*. Amsterdam: Elsevier/North-Holland, 417–554.

Laudan, L. (1981), A confutation of convergent realism, *Philosophy of Science* **48**: 19–49.

Leopold, J. and I. Percival (1980), The semiclassical two-electron atom and the old quantum theory, *Journal of Physics B: Atomic and Molecular Physics*, **13**: 1037–47.

Liboff, R. (1984), The correspondence principle revisited, *Physics Today* **37**: 50–5.
Lu, K.T., F.S. Tomkins, and W.R.S. Garton (1978), Configuration interaction effect on diamagnetic phenomena in atoms: strong mixing and Landau regions, *Proceedings of the Royal Society of London A*, **362**: 421–4.
Main, J., G. Weibusch, A. Holle, and K.H. Welge (1986), New quasi-Landau structure of highly excited atoms: the hydrogen atom, *Physical Review Letters*, **57**: 2789–92.
Massimi, M. (2005), *Pauli's Exclusion Principle: The Origin and Validation of a Scientific Principle*. Cambridge: Cambridge University Press.
Matzkin, A. (2006), Can Bohmian trajectories account for quantum recurrences having classical periodicities?, arXiv.org, preprint quant-ph/0607095.
McAllister, J. (1996), *Beauty and Revolution in Science*. Ithaca, NY: Cornell University Press.
McMullin, E. (1978), Structural explanation, *American Philosophical Quarterly* **15**: 139–47.
McMullin, E. (1984), A case for scientific realism, in Jarrett Leplin (ed.) *Scientific Realism*. Berkeley, CA: University of California Press, pp. 8–40.
McMullin, E. (1985), Galilean idealization, *Studies in History and Philosophy of Science* **16**: 247–73.
Mehra, J. and H. Rechenberg (1982), *The Historical Development of Quantum Theory: Volume I: The Quantum Theory of Planck, Einstein, Bohr and Sommerfeld: Its Foundation and the Rise of Its Difficulties 1900–1925, & Volume II: The Discovery of Quantum Mechanics 1925*. New York: Springer-Verlag.
Meredith, D. (1992), Semiclassical wavefunctions of nonintegrable systems and localization on periodic orbits, *Journal of Statistical Physics* **68** (1/2): 97–130.
Messiah, A. (1965), *Quantum Mechanics*, vols. I and II. G.M. Temmer (trans.). Amsterdam: North-Holland Publishing.
Mitchell, S. (2003), *Biological Complexity and Integrative Pluralism*. Cambridge: Cambridge University Press.
Morrison, M. (1999), Models as autonomous agents, in M. Morgan and M. Morrison (eds.), *Models as Mediators: Perspectives on Natural and Social Science*. Cambridge: Cambridge University Press: 38–65.
Nagel, E. ([1961] 1979), *The Structure of Science: Problems in the Logic of Scientific Explanation*. Indianapolis: Hackett Publishing.
Narimanov, E.E., N.R. Cerruti, H.U. Baranger, and S. Tomsovic (1999), Chaos in quantum dots: Dynamical modulation of Coulomb blockade peak heights, *Physical Review Letters*, **83**: 2540–643.
Narimanov, E.E., H.U. Baranger, N.R. Cerruti, and S. Tomsovic (2001), Semiclassical theory of Coulomb blockade peak heights in chaotic quantum dots, *Physical Review B*, **64**: 235329–1 to 235329–13.
Nielsen, J.R. (1976), Introduction to *Niels Bohr Collected Works, Vol. 3*, in Bohr (1976), pp. 3–46.
Nickles, T. (1973), Two concepts of intertheoretic reduction, *The Journal of Philosophy* **70**: 181–201.
O'Connor, P., J. Gehlen, and E. Heller (1987), Properties of random superpositions of plane waves, *Physical Review Letters*, **58**: 1296–9.
Oppenheim, P. and H. Putnam (1958), Unity of science as a working hypothesis, in H. Feigl, G. Maxwell, and M. Scriven (eds.) *Minnesota Studies in the Philosophy of Science*, Vol 2. Minneapolis: University of Minnesota Press.
Park, H-K., and S.W. Kim (2003), Decoherence from chaotic internal dynamics in two coupled δ-function-kicked rotors, *Physical Review A* **67**: 060102–1 to 060102–4.

Parker, W. S. (forthcoming), Does matter really matter? Computer simulations, experiments, and materiality, *Synthese*.
Pauli, W. (1926), Quantentheorie, in H. Geiger and K. Scheel (eds.), *Handbuch der Physik*, 23: 1–278.
Peres, A. (1995), *Quantum Theory: Concepts and Methods*. Dordrecht: Kluwer Academic.
Planck, M. (1906), *Vorlesungen uber die Theorie der Warmestrahlung*. Leipzig: Johann Ambrosious Barth.
Planck, M. (1913), *Vorlesungen uber die Theorie der Warmestrahlung* , 2nd edition. Leipzig: Johann Ambrosious Barth.
Planck, M. (1959), *Theory of Heat Radiation*. Translation of Planck (1913). New York: Dover.
Popper, K. ([1963] 1989), *Conjectures and Refutations: The Growth of Scientific Knowledge*. London: Routledge.
Post, H.R. (1971), Correspondence, invariance and heuristics: In praise of conservative induction, *Studies in the History and Philosophy of Science* **2**: 213–55.
Putnam, H. (1975), *Mathematics, Matter and Method. Philosophical Papers*, vol. 1. Cambridge: Cambridge University Press.
Radder, H. (1991), Heuristics and the generalized correspondence principle, *British Journal for the Philosophy of Science* **42**: 195–226.
Railton, P. (1980), *Explaining Explanation: A Realist Account of Scientific Explanation and Understanding*. Ph.D. Dissertation, Princeton University.
Redhead, M. (2001a), The intelligibility of the universe, in A. O'Hear (ed.) *Philosophy at the New Millennium*. Cambridge: Cambridge University Press, pp. 73–90.
Redhead, M. (2001b), Quests of a realist: Review of Stathis Psillos, *Scientific Realism*, *Metascience* **10**: 341–7.
Redhead, M. (2004), Discussion note: Asymptotic reasoning, *Studies in History and Philosophy of Modern Physics* **35**: 527–30.
Richter, K. (2000), *Semiclassical Theory of Mesoscopic Quantum Systems*. New York: Springer-Verlag.
Robnik, M. (1981), Hydrogen atom in a strong magnetic field: On the existence of the third integral of motion, *Journal of Physics A* **14**: 3195–216.
Rosenfeld, L. ([1973] 1979), The wave-particle dilemma, in R. Cohen and J. Stachel (eds.) *Selected Papers of Léon Rosenfeld*. Boston Studies in the Philosophy of Science 21. Dordrecht: D. Reidel Publishing Co., 688–703. Originally published in J. Mehra (ed.) *The Physicist's Conception of Nature*. Dordrecht: D. Reidel Publishing Co., 251–63.
Salmon, W. (1984), *Scientific Explanation and the Causal Structure of the World*. Princeton: Princeton University Press.
Salmon, W. (1989), Four decades of scientific explanation, in P. Kitcher and W. Salmon (eds.) *Scientific Explanation: Minnesota Studies in the Philosophy of Science* 13.
Sánchez-Ron, J. (1983). Quantum vs. classical physics: Some historical considerations on the role played by the "principle of correspondence" in the development of classical physics, *Fundamenta Scientia* **4**: 77–86.
Sarkar, S. (1992), Models of reduction and categories of reductionism, *Synthese* **91**: 167–94.
Saunders, S. (1993), To what physics corresponds, in S. French and H. Kaminga, (eds.) *Correspondence, Invariance, and Heuristics; Essays in Honour of Heinz Post*. Dordrecht: Kluwer Academic, pp. 295–326.
Scerri, E. (1994), Has chemistry at least approximately been reduced to quantum mechanics?, *PSA Proceedings of the Biennial Meeting of the Philosophy of Science Association 1994*, Vol. **1**: Contributed Papers, 160–70.
Schaffner, K. (1967), Approaches to reduction, *Philosophy of Science* **34**: 137–47.

Scheibe, E. (1988), The physicists' conception of progress, *Studies in History and Philosophy of Science* **19**: 141–59. Reprinted in Scheibe (2001), 90–107.
Scheibe, E. (1993), Heisenberg's concept of a closed theory. In Scheibe (2001), 136–41.
Scheibe, E. (2001). *Between Rationalism and Empiricism: Selected Papers in the Philosophy of Physics*. Edited by B. Falkenburg. New York: Springer Verlag.
Schlosshauer, M. (2006), Comment on "Quantum mechanics of Hyperion", Preprint quant-ph/0605249.
Schulman, L. S. (2005), *Techniques and Applications of Path Integration*. New York: Dover.
Schwarzschild, K. (1916), Zur Quantenhypothese, *Berliner Berichte 1916*: 548–68.
Schweber, S. S. (1994), *QED and the Men Who Made It*. Princeton: Princeton University Press.
Sklar, L. (1999), The reduction(?) of thermodynamics to statistical mechanics, *Philosophical Studies*: **95**: 187–202.
Sommerfeld, A. (1916), Zur Quantentheorie der Spektrallinien, *Annalen der Physik*, **51**: 1–94; 125–67.
Sommerfeld, A. ([1919] 1923), *Atomic Structure and Spectral Lines*. Translated by H. Brose. London: Methuen.
Sommerfeld, A. (1920), Letter to Niels Bohr, November 1920. In Bohr (1976) p. 690.
Suppe, F. (1989), *The Semantic Conception of Theories and Scientific Realism*. Urbana: University of Illinois Press.
Sridhar, S. (1991), Experimental observation of scarred eigenfunctions of chaotic microwave cavities, *Physical Review Letters* **67**: 785–8.
Stone, A. D. (2005), Einstein's unknown insight and the problem of quantizing chaos, *Physics Today*, August: 1–7.
Stoppard, T. (1988), *Hapgood*. Boston: Faber and Faber.
Styer, D., M. Balkin, K. Becker, *et al.* (2002), Nine formulations of quantum mechanics, *American Journal of Physics* **70**(3): 289–97.
Tabor, M. (1989), *Chaos and Integrability in Nonlinear Dynamics: An Introduction*. New York: Wiley Interscience.
Tanner, G., K. Richter, and J.-M. Rost (2000), The theory of two-electron atoms: Between ground state and complete fragmentation, *Reviews of Modern Physics* **72**: 497–544.
Tanona, S. (2002), From correspondence to complementarity: The emergence of Bohr's Copenhagen interpretation of quantum mechanics, Ph.D. Dissertation, Indiana University.
Tanona, S. (2004), Uncertainty in Bohr's response to the Heisenberg microscope, *Studies in History and Philosophy of Modern Physics* **35**: 483–507.
Thomson-Jones, M. (2006), Models and the semantic view, *Philosophy of Science* **73**: 524–35.
Trout, J. D. (2002), Scientific explanation and the sense of understanding, *Philosophy of Science* **69**: 212–33.
Uzer, T., D. Farrelly, J. A. Milligan, P. E. Raines, and J. P. Skelton (1991), "Celestial mechanics on a microscopic scale", *Science*, **253**: 42–8.
van der Waerden, B. (ed.) (1967), *Sources of Quantum Mechanics*. New York: Dover Publications.
van Fraassen, B. (1980), *The Scientific Image*. Oxford: Oxford University Press.
van Fraassen, B. (1997), Structure and perspective: Philosophical perplexity and paradox, in M. L. dalla Chiara, K. Doets, D. Mundici, and J. van Benthem (eds.) *Logic and Scientific Methods*, Volume One of the Tenth International Congress of Logic, Methodology and Philosophy of Science, Florence, August 1995. Dordrecht: Kluwer Academic.

van Fraassen, B. (2006), Structure: Its shadow and substance, *British Journal for the Philosophy of Science* **57**: 275–307.
van Vleck, J. (1922), The dilemma of the helium atom, *Physical Review* **19**: 419–20.
von Baeyer, H.C. (1995), The philosopher's atom, *Discover* **16** (11).
Weinberg, S. (1992), *Dreams of a Final Theory: The Scientist's Search for the Ultimate Laws of Nature*. London: Trafalgar Square Publishing.
Weinert, F. (1994), The correspondence principle and the closure of theories, *Erkenntnis* **40**: 303–23.
Welch, G.R., M.M. Kash, C. Iu, L. Hsu, and D. Kleppner (1989), Positive-energy structure of the diamagnetic Rydberg spectrum, *Physical Review Letters* **62**: 1975–8.
Wentzel, G. (1926). Eine Verallgemeinerung der Quantenbedingung fur die Zwecke der Wellenmechanik, *Zeitschrift fur Physik* **38**: 518–29.
Wiebe, N. and L.E. Ballentine (2005), Quantum mechanics of Hyperion, *Physical Review A* **72**: 022109–1 to 022109–15.
Wigner, E.P. (1932), On the quantum correction for thermodynamic equilibrium, *Physical Review* **40**: 749–59.
Wigner, E.P. (1960), The unreasonable effectiveness of mathematics in the natural sciences, *Communications in Pure and Applied Mathematics* **13**: 1–14.
Wimsatt, W.W. (1987), False models as means to truer theories, in M. Nitecki and A. Hoffman (eds.) *Neutral Models in Biology*. Oxford: Oxford University Press.
Wintgen, D., K. Richter, and G. Tanner (1992), The semiclassical helium atom, *Chaos*, **2**: 19–33.
Wise, M.N. and D.C. Brock (1998), The culture of quantum chaos, *Studies in History and Philosophy of Modern Physics* **29**: 369–89.
Wisdom, J., S. Peale, F. Mignard (1984), The chaotic rotation of Hyperion, *Icarus* **58**: 137–52.
Woodward, J. (2003), *Making Things Happen: A Theory of Causal Explanation*. Oxford: Oxford University Press.
Woodward, J. and C. Hitchcock (2003), Explanatory generalizations, part I: A counterfactual account, *Noûs* **37**: 1–24.
Worrall, J. (1989), Structural realism: The best of both worlds?, *Dialectica* **43**: 99–124.
Young, L. and G. Uhlenbeck (1930), On the Wentzel-Brillouin-Kramers approximate solution of the wave equation, *Physical Review* **36**: 1154–67.
Zurek, W. (1991), Decoherence and the transition from quantum to classical, *Physics Today* **44**: 36–44.
Zurek, W. (1998), Decoherence, chaos, quantum-classical correspondence, and the algorithmic arrow of time, *Physica Scripta* T**76**: 186–98.
Zurek, W. (2003), Decoherence, einselection, and the quantum origins of the classical, *Reviews of Modern Physics* **75**: 715–75.
Zurek, W. and J.P. Paz (1997), Why we don't need quantum planetary dynamics: Decoherence and the correspondence principle for chaotic systems, in B.L. Hu and D.H. Feng (eds.) *Quantum Classical Correspondence: Proceedings of the 4th Drexel Symposium on Quantum Nonintegrability, Drexel University, September 8–11, 1994*. Cambridge, MA: International Press, pp. 367–79.

Index